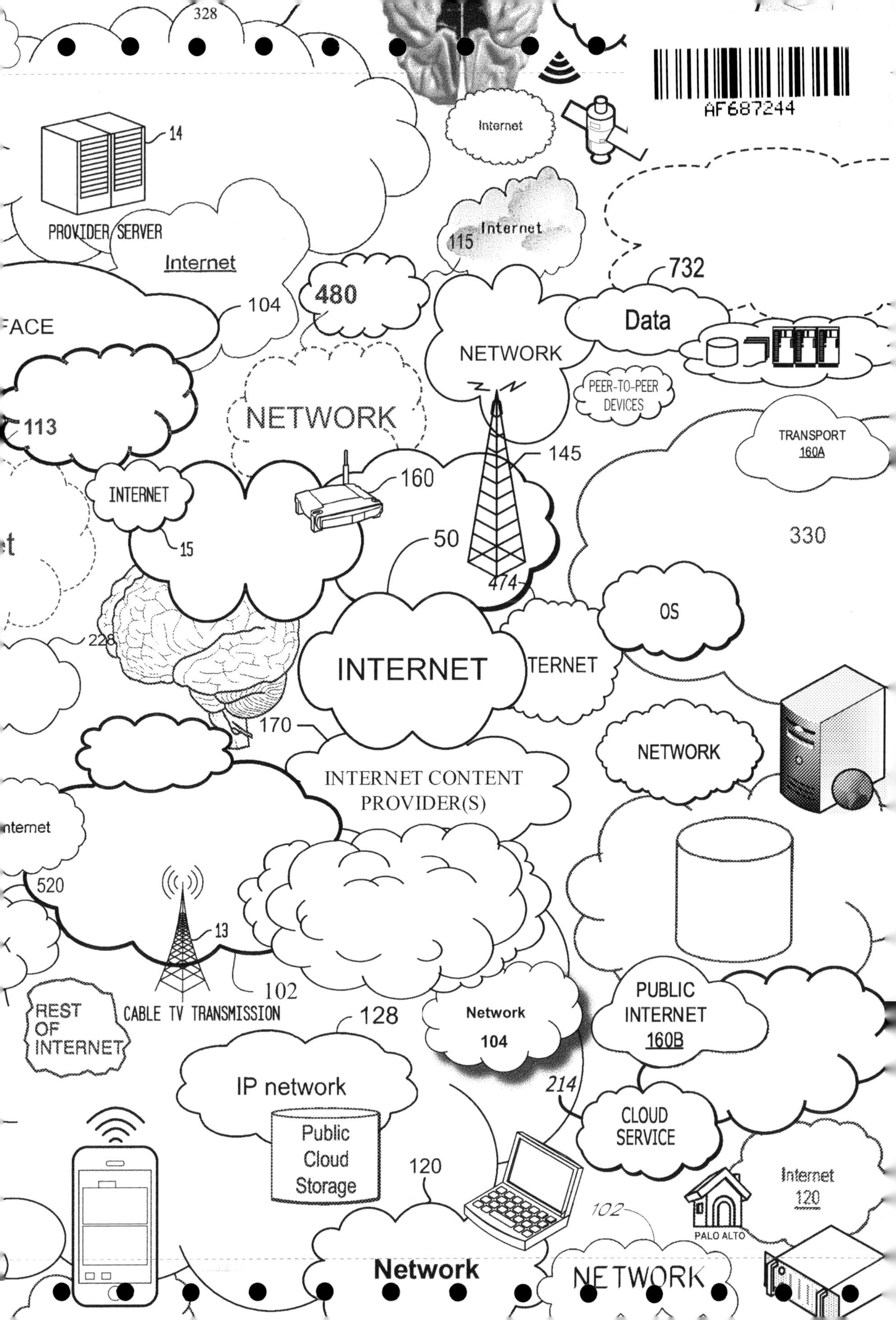
328
AF687244
14
PROVIDER SERVER
Internet
Internet
115
Internet
732
104
480
Data
ACE
NETWORK
NETWORK
PEER-TO-PEER
DEVICES
113
TRANSPORT
160A
145
160
INTERNET
50
15
330
474
OS
228
INTERNET
TERNET
170
NETWORK
INTERNET CONTENT
PROVIDER(S)
nternet
520
13
102
REST
OF
INTERNET
CABLE TV TRANSMISSION
128
Network
104
PUBLIC
INTERNET
160B
IP network
214
Public
Cloud
Storage
CLOUD
SERVICE
120
Internet
120
102
PALO ALTO
Network
NETWORK

Peter Schrader
1936 – 2020

ACID CLOUDS

MAPPING DATA CENTRE TOPOLOGIES

Edited by Niels Schrader and Jorinde Seijdel
With photographs by Roel Backaert

nai010 publishers

Prologue
Acid Clouds: Mapping Data Centre Topologies

Niels Schrader and Jorinde Seijdel

> **If data centres crystallise political powers, they are likely to become a crucial point of investigation and intervention for movements that seek to change the world.**
> – Brett Neilson and Ned Rossiter, *Automating Labour and the Spatial Politics of Data Centre Technologies*, 2022

The 'becoming' of *Acid Clouds: Mapping Data Centre Topologies* has been marked and stretched by multiple Covid-19 interferences and global lockdowns. Before materializing into a book, it took many forms, from artistic research at the Royal Academy of Art in The Hague to photographs, video interviews, 1 data centre walks, 2 exhibitions, workshops, lectures and seminars. 3 Niels Schrader, information designer and co-editor of this book, initiated *Acid Clouds* as a project. Collaborating with photographer Roel Backaert, he aimed to map the material traces of virtual data in the Netherlands and question the prevailing image of 'the cloud' as a sterile and environmentally friendly domain. Starting in 2018, Schrader and Backaert systematically captured Dutch data centres at night. Donned in black, the photographers blended in with the highly secured dark environment. They meticulously documented almost two hundred, revealing a post-industrial typology and distinct visibility to its presence as an object, location and concept.

The in-depth interviews Schrader conducted with experts from the Dutch Internet industry, tech activism and technology critique unveil the complex impact of data centres on their immediate environment and society at large. These conversations underscore that data centres as storage facilities raise pressing ethical and political questions for ecology and sustainability, human rights and equality, data ownership and control, urban development and social cohesion, capitalist expansion and extractivism, as well as artificial intelligence and machine learning. Through these it became evident that most data centres cannot be studied merely as local phenomena, but must be seen as integral parts of the global network infrastructures.

A plan was conceived to delve deeper into these topics through a book featuring essays, photography, a technical glossary, diagrams, graphs and maps. Editor, writer and teacher Jorinde Seijdel who has a keen interest in the cultural implications of media and technology, joined as co-editor to collaborate with nai010 publishers. The objective was to outline various data centre topologies and provide a diverse range of perspectives with international guest authors. *Acid Clouds: Mapping Data Centre Topologies* includes contributions by Ramon Amaro and Sheena Calvert, Annet Dekker, Mél Hogan, Michiel van Iersel and Mark Minkjan with drawings by Anne Dessing, Ned Rossiter, Niels Schrader, Jorinde Seijdel with illustrations by Alice Mulder, Marina Otero Verzier, Max Kortlander, Marleen Stikker and Sander van der Waal from Waag, and Füsun Türetken. The book is divided into six chapters, five of which featuring a brief editorial introduction – 'Governance', 'Materialities', 'Aesthetic Energies', 'Liberating Identities', and 'Mapping Acid Clouds' – and a closing 'Appendix' with technical information.

'Governance' critically questions the power constellations within which data centres operate. Media theorist Ned Rossiter explores the global infrastructures of digital capitalism, emphasizing the need for reappropriation and the role of counter-imaginaries. Max Kortlander, Marleen Stikker and Sander van der Waal from Waag in Amsterdam, advocate for transparency, public values, and the common good, addressing the rights and responsibilities involved in the interaction between data centres and society. In particular, they stress the importance of technology design and implementation processes.

The second section 'Materialities' focusses on the physical manifestations of the cloud, where Schrader looks into the impact of resources needed to create, share and store daily virtual data output with respect to the physical environment. Artist and academic Füsun Türetken examines the role of metal in shaping the world of capitalist finance, belief systems, geopolitical relations, and (digital) bodies, arguing that capitalism leaves a planet suffering from 'Gaiaporosis'.

The emphasis shifts to the artistic and cultural perceptions of data centres in 'Aesthetic Energies'. Annet Dekker, media theorist and curator, explores the socio-technical entanglements and ethical implications of data centres through the analysis of Caleb Larsen's artwork *A Tool to Deceive and Slaughter* (2009). Seijdel presents a glossary containing terms and references forming an affective and critical network around the data centre rather than serving it. Urbanists Michiel van Iersel and Mark Minkjan conclude this section in evaluating the urban and social history of data centres from a futuristic perspective, imagining the decentralization of the Internet and the re-politicization of data.

The last section 'Liberating Identities' examines, from various perspectives, the oppression, inequality and exploitation resulting from imbalanced power relations in the production and control of data and technology. Mél Hogan, director of Canada's Environmental Media Lab, focusses on the computational power enabled by the data centre as a sign of white settler scientific progress and asks which kind of human is created and reflected by AI. Ramon Amaro, Senior Researcher in Digital Culture at Nieuwe Instituut in Rotterdam, and philosopher/designer Sheena Calvert discuss the complex relationship between racism and machine learning, drawing on the history of data and statistical analysis. Finally, architect and researcher Marina Otero Verzier rethinks the legitimacy of data-driven, extractive societies by engaging with the Atacameño communities in Chile fighting lithium mining in their salt flats.

While the title of the *Acid Clouds* project initially referred mainly to the notion of acid rain and its mineral composition due to technology and the quasi-innocent nature of the name 'cloud' computing, in the book it came to work as a metaphor for our current digital condition. This publication unfolds diverse data centre topologies, which are intertwined and interconnected, and with the 'acid clouds' taking on multiple meanings: they shift and overlap across social, cultural, political and geographical fields. What becomes evident in each contribution is that the growing ubiquity of data, algorithms and machine learning in daily life, the workplace, and communities requires a more proactive and empowering attitude. Data centres must be questioned from the intersections of body, identity, digital technology, artistic expression and activism. This prologue closes with several such questions important now, and in the future: How can one become more aware of the effects of digital habits on the world? How can one reimagine a relationship with data? How can one foster diverse understandings of technology and resist the extractive logic of contemporary data regimes?

The editors and Roel Backaert express their sincere gratitude to the following individuals and institutions for their support, interest, advice, generosity and patience in conceiving and realizing this book: the authors, Janine Armin, Giulia Cundari, Michał Ejdys, Geert Lovink, Lectorate Design at the Royal Academy of Art in The Hague, Waag in Amsterdam and -1 at the Nieuwe Instituut in Rotterdam.

1. To identify different types of problems connected to data centres, Schrader has conducted video interviews with industry leaders, climate activists, advocates for data privacy, media theorists and members of local action groups. The interviews were held from 30 January 2019 to 26 September 2022, and their job descriptions at the time of the interviews are noted here: Ramon Amaro (Lecturer in Art and Visual Cultures of the Global South, University College London), Michel van den Assem (Managing Director, Interxion), Evelyn Austin (Executive Director, Bits of Freedom), Stijn Grove (Managing Director, Dutch Data Center Association), Walter Hoogland (former Scientific Director, Nikhef and former Research Director, CERN), Martin Lloyd (Mobilisation and Communication Director, Greenpeace Netherlands), Gerard Rosbach (architect, Rosbach Architects), Lars Ruiter (Municipality Councillor Hollands Kroon and activist) and Marleen Stikker (founder, Waag). See the *Acid Clouds* website for abstracts of some recordings. Online. Available at *www.acidclouds.org* (accessed 10 April 2024).
2. As part of Waag, *Expedition: Future*, Amsterdam, 23–28 September 2021.
3. *Acid Clouds* as a project was presented at, amongst others, Elisava – Barcelona School of Design and Engineering, Independent School for the City, (Rotterdam), Framer Framed (Amsterdam), Gerrit Rietveld Academie (Amsterdam) and Delft University of Technology.

Atlas international des nuages
(International Cloud Atlas),
Hugo Hildebrand Hildebrandsson, 1896.

V

ALTO-CUMULUS

Fig. 9.

ALTO-CUMULUS

Fig. 10.

Photochromotypie Brunner & Hauser, Zürich.

X

Cumulus

Fig. 19.

Cumulus

Fig. 20.

Photochromotypie Brunner & Hauser, Zürich.

Atlas international des nuages
(International Cloud Atlas),
Hugo Hildebrand Hildebrandsson, 1896.

XI

FRACTO-CUMULUS

Fig. 21.

CUMULO-NIMBUS

Fig. 22.

Photochromotypie Brunner & Hauser, Zürich.

Atlas international des nuages
(International Cloud Atlas),
Hugo Hildebrand Hildebrandsson, 1896.

The present internationally adopted system of cloud classification is the result of work that started in the nineteenth century. The first attempt to use photography for cloud classification was made by Hugo Hildebrand Hildebrandsson, in Uppsala, who prepared a cloud atlas of 16 photographs. The further development of this work resulted in the publication in 1896 of the first International Cloud Atlas, containing 28 coloured plates accompanied by definitions and descriptions of clouds and instructions on cloud observations in French, German, and English.

Governance – The Cloud's Spatial Strategies

This chapter is a critical interrogation of the local and global power constellations of which data centres are a crucial part. From different perspectives and at different levels it examines how data centres relate to the public, the private and the political through both established and missing relationships with state agencies, digital rights organizations, citizens' initiatives and Big Tech. What could be changed about existing relationships and structures to achieve a more equitable exchange of data ownership and control? And, what are the responsibilities of the data centre industries, their lobby organizations, the governments, citizens and individual users?

Ned Rossiter, a media theorist known for his research on network cultures, the politics of cultural labour, logistical media and data politics, focusses explicitly on the global infrastructures of digital capitalism. He emphasizes the need to re-appropriate technological infrastructures and points to the potentially liberating role of counter-imaginaries in the political geography of data. According to Rossiter, a key step in this process is to identify the dominant, black box control systems and collectively devise ways of exploiting these apparatuses from within.

Marleen Stikker's publication *The Internet Is Broken, but We Can Fix It*, 1 lays the foundations for her contribution in collaboration with Max Kortlander and Sander van der Waal. From the perspective that the Internet is not yet beyond repair, Stikker, Kortlander and Van der Waal discuss the rights and obligations related to the design, construction and operation of data centre facilities. In defence of transparency and public values, they question the role of data centres and propose instead the model of a 'public stack': a collection of open, fair and secure alternatives overseen by the commons – including devices, applications, processes, relationships – that guarantee the public interest and can be placed at the centre of design and implementation processes of technology.

1. M. Stikker, *Het internet is stuk. Maar we kunnen het repareren* (Amsterdam: De Geus, 2019).

Clouds of Power and the Politics of Data

Ned Rossiter

> **Sovereignty ... may have found new life in the data bunker.**
> – Tung-Hui Hu, *A Prehistory of the Cloud*, 2015

> **A new geopolitics appears to emerge from this need for cooling ... the internet is being built with a specific natural geography and climate in mind.**
> – Nicole Starosielski, *Media Hot & Cold*, 2021

While corridors of power are defined as the scene and site where the most important decisions are made, the palace of the digital age is the data centre. Within these data facilities, corridors are arteries for the transmission of thermal conditions. Corridors govern movement. Considered a key index of architectural modernity, corridors regulate the circulation of people and things across a range of building types. With the design and construction of data centres, however, the flow of humans and objects becomes an exception, while the circulation of air is critical for the optimization of machines and security of data. The separation of cold and hot air through distinct corridor systems has become the standard architectural design for data centres, essentially serving the servers with cool air by extracting hot air down the corridor. In the case of the data centre, corridors divide and disconnect distinct areas into different climatic zones within the building in the endeavour to further economize exorbitant energy use for cooling data centres. At the same time, the separation of climatic zones allows for a more efficient monitoring of risks, such as hot spots. The building type of the data centre is considered increasingly as a machine, a twenty-first-century mega-version of the 1960s mainframe, one giant computer with the design logic of the corridor.

As infrastructural monuments of digital capitalism, data centres facilitate internal east-west traffic (between servers in the same facility), which exceeds external north-south traffic (between servers and clients beyond the data centre). The primacy of the east-west axis signals how the concentration or clustering of businesses in the same installation enables competitive advantage over the north-south axis. In the case of the former, such circuits of transaction are made possible by peering relations and data connections between firms stationed in different virtual machines on the same or adjacent server racks. Consisting of hardware and software, networked traffic passes through patched cables that usually intersect in a meet-me room. These cables follow different paths than the hot/cold corridors. Computational decisions to connect are motivated by operational advantages obtained by low latency, accessibility, power consumption and scalability. [1]

At the time of writing there are 4,913 colocation data centres in 130 countries around the world, if we are to take the figures from the Colocation Data Centers site as an approximation. [2] The United States commands 1,831 of these, the United Kingdom has 268 and Germany is host to 233. Australia has 129, India 162 while Hong Kong, Indonesia, and Taiwan hold 59, 53 and 10 respectively. China has 87, Russia 59, Ireland 27 with Luxembourg hosting 15, the Czech Republic 24, Malaysia 32, South Africa 26, Pakistan has 19 and Turkey 65. The Netherlands has 199 and Ukraine 34. This represents a 30–50 per cent growth in the sector over the past 5 to 7 years. With their capacity for supporting 'peering arrangements' that enable faster exchanges of data, colocation data centres are primarily geared towards finance capital. However, there is no obvious reason why finance capital should hold a monopoly of interest in colocation data centres given the types of service offered at many of these centres. These services include dedicated servers, virtual servers, cages, rack cabinets, managed hosting and so forth – none of which are exclusive to high-frequency trading and the competitive advantage obtained with optimized low latency (fast connections), but are features also of data economies more generally. This essay surveys some of the primary political issues that attend the burgeoning data economies and the arguably short-term business horizon of data centres.

Data Centres as Infrastructures of Circulation

The global networks of supply chains expand the territorial reach of producers and suppliers required for the operation of the logistical state. Enterprise resource planning (ERP) software systems calibrate labour productivity and coordinate the movement of goods and finance in real-time. Data centres store, process and transmit the data integral to logistical operations governed by computational systems. Combined, these software and infrastructural components make possible imperial rule for the logistical state. Importantly, the spatial and temporal dimensions described here are not synchronic or spatial equivalents. Time and space is peculiar to each, forming layers or, more likely, a complex undulation of planes that overlap and intersect on some occasions while colliding and disconnecting on others. Yet we know all too well the materiality of the digital: from the monocrystalline silicon substrate of printed circuit boards to the architectural form of data centres stuffed with server racks, from the copper alloy of coaxial cable to the bodies in pain that mine elemental metals such as copper from the Chilean mountains. As a technological object and electronic system, the digital is both produced and conditioned by multiple variants of the material world.

Like much of the infrastructure that supports our daily life and global economies, the topic of data centres is one that is largely shrouded in mystery. The very banality of communication rarely compels our imagination to take the step into making the concrete abstraction of infrastructure something we might know and address. The rise of data centres integrates society with an economy whose technical infrastructure is defined by storage, transmission and processing. The data centre industry has been accelerated by the proliferation of digital media throughout social life. As large-scale digital infrastructures, data centres are made operational on a planetary scale in the transition from mainframe computing to facilities able to accommodate multiple servers that support the digital economy of software-as-a-service (SaaS), a term that refers to the delivery of software as a series of features, or 'services', over a network rather than as an executable file that installs and runs from a computer's hard disk.[3] Think cloud computing or the low latency networks required for high-frequency trading followed by other sectors further down the food chain of speed (military-industrial complexes, medical services, urban systems and entertainment industries). But also consider the way institutional diversity is increasingly hard to identify as a result of the limited parametric variation of combinatory options availed by software services to any organization – logging, billing, visualization, data authentication, predictive analytics, business intelligence, search, conversion, publication and backup. Operating outside such limits and inventing new systems of organization and cultures of expression involves a parametric politics of the present. Arguably, the problematic of engineers determining the design of operational borders (parametric politics) is less about infrastructure and more about the cultivation and interjection of atmospheres – that is, the creation of working environments, decision-making processes and affective spaces not entirely governed by digital systems. Such offline efforts register how atmospheres are also key to the infrastructural and medial condition of organizational cultures and practices. Let's keep in mind that such a move occurs within a context of data analytics and a certain neo-positivist epistemological and institutional anxiety around the production of 'evidence-based' research and the economies it fuels. In other words, there is more to empirical conditions and knowledge production than accounts of the world generated by computational systems and data analytics. In probing the question of infrastructures and their atmospheres, we might ask where the data centres are located and whether this has any bearing on the design of business imaginaries of cloud-based economies. Indeed, what sort of data traffics through different data centres distributed in strategic locations throughout the world? Are there juridical regimes specific to different types of data? How is the provenance of data complicated in legal ways by the location of its storage? What sort of protocols of hardware and storage are required for, say, financial data as distinct from the data collected by the state or military apparatus? Can communication protocols also be produced from what Soenke Zehle terms the 'logistical lifestreams' that define our social media worlds? 4 And to what extent do the technical operations, geography of location, and political economy specific to different data centres determine the types of businesses and organizational practices dependent on hosting services?

These are questions that also go some way towards a media theory of data centres, which are a key communication infrastructure for the logistical processes of contemporary capitalism and global supply chains. Taiwan, Hong Kong and Singapore are strategic sites for investment in storage facilities for cloud computing. The past 10 years or so has seen a rapid growth in the construction of data centres around the world and within the Asian region. Such developments signal a historic and paradigmatic shift from the extraction of surplus value from labour as a core structural dynamic of the capitalist mode of production to an arguably novel form of surplus information within an economy of algorithmic capitalism. In the first instance, this new form of surplus seems less dependent on labour-power and instead generated by machines inasmuch as data centres are largely facilities depopulated of workers, aside from a handful of security guards, technicians and site managers. The conventional Marxian labour theory of value, which ties the value of commodities to the production cost of labour, might necessarily extend to or be complemented by a machinic theory of value.

This is especially the case with high-frequency trading, where the nano-speed of trade operates beyond the threshold of perception and with minimal human oversight or intervention. But the infrastructure of data centres, when understood beyond the technical properties of infrastructure, starts to make clear that labour remains not only as a condition of possibility, but also a necessary supplement to the algorithm, filling in the gaps for processes not yet automated by the machine. Amazon's Mechanical Turk is one chief example here, where Human Intelligence Tasks (HITs) are enlisted for low value services of microwork not yet accommodated by algorithms. If the future of work scrapes the dull surface of the mind, what lines of the political might be staged against infrastructures that precipitate the obsolescence of labour and not, simply, its transformation into new jobs, as narratives of industrial modernity would have it?

This question invites a revisiting of theories of the state as a sovereign entity that prepares the ground for the construction and commercial viability of data centres. How might we revive a critique of the state based on its support of commercial data centres through financial levers such as attractive corporate tax breaks, generous land leases, flexible labour regulations and free trade zones among other forms of state subsidy funded by national citizenries? Another important consideration in analysing the nexus between state formation and digital infrastructure includes lines of revenue generated by the work of non-citizens residing in the territory of the state. This is a feature of the platform economies and their frequent dependency on 'gig workers', many of whom reside in national settings without secure residency. Certainly, there has long been a crucial labour cost related to questions of infrastructure and geography, as we can note in the protracted struggle by German labour union Verdi in its demands for better working conditions in Amazon's warehouses. [5]

Related to the geography of data centres are the imperial and colonial histories of communications infrastructure – the submarine telegraph cables that first established the corridors of trade, connecting the territory of empire to the speed of communication. [6] Consider, also, the emergence of what is referred to in one higher education business report as 'the financially sustainable university', which in seeking to overcome the fragmentation of data-driven processes due to the conflicts and complexities of internal operations, advises instead the use of outsourcing data centres: 'Third-party data centres, whether they are managed or cloud-based, could provide more sophisticated solutions, higher levels of security, greater flexibility in capacity and lower cost than internal solutions – all with greater accountability and less politics'. [7] In striving after evermore enhanced techniques of optimization and efficiency, the logistical university becomes indistinct from any number of organizations whose core goal is to extract maximal value from data-driven processes.

Data Centres in the Age of Cyberwarfare

Data centres are not the exclusive domain of data that matters in the murky theatre of war and military atrocity meted out to civilian populations. [8] Word of mouth, rumours, disinformation, propaganda, fake news, the cacophony of noise – all come to bear as orientation devices and resources for action. None are exclusive to data centres. Needless to say, digitally encoded information is key to the determination of attack and counter-attack. Arial surveillance units using drones and satellites, encrypted communications and cloud-based management systems are among the logistical media whose operational capacity is tied to data centres. [9]

So, what analytical status might we accord to data centres in such contexts and conditions? What do data sanctions mean in the context of war? And how do data centres figure as infrastructures of imperialism in a conflict predicated on territorial expansion by an aggressor state? Does the private tech sector in the business of data centres and cloud services function as a counter-imperial force or rather as a partner in crime? And how do artificial intelligence and machine learning technologies assume decision-making tasks related to the orchestration, governance and oversight of operations in times of military conflict? What does a data centre do in times of war? These are framing questions that might orientate ongoing research into the geopolitics and geoeconomics of data centres.

Within data economies and computational cultures of real-time, automated and human decision-making is coincident with the traffic of data through hardware circuits and networked digital infrastructure. But how does a parallel temporality of standby, or what literary theorist and engineer Tung-Hui Hu calls 'wait time', unsettle or disrupt human expectations and machine operations tied to immediacy? [10] Drawing on Paul Virilio's concept and study of 'bunker archaeology', Hu considers bunkers as 'fundamentally a *temporal* architecture' that 'produces the imagination of a future looking backward on the current moment and, as such, incorporates a phantasmic time outside the boundaries of lived time'. [11] For Hu, digital or data bunkers also carry these features. [12] If, as Hu notes, cloud security systems produce an atmosphere and 'bunker mentality' of 'passive watchfulness' and fear-infused vigilance, then cloud-based software programs take command of computational decisions whose potential consequences are always already immanent to real-time operations. Microsoft's cloud-based security information and event management system, Sentinel, embodies these sort of attributes in both name and description, promoted for its 'birds-eye view across the enterprise alleviating the stress of increasingly sophisticated attacks, increasing volumes of alerts, and long resolution time frames'. [13] How, then, might the multiple temporalities of cloud infrastructures and data centres structure and organize cultures of perception and modes of action and decision in the midst of military incursion that seeks to destroy or defend sovereign integrity? Indeed, are data sovereign in ways that undermine the monopoly of violence assumed of the modern state?

The 'data sovereignty' debate often circles around notions that reclaim the policy territory for 'national' IT solutions. Yet, as Primavera De Filippi and Smári McCarthy note, 'Data, as such, does not have any nationality but merely inherits the law of the territory in which it is located.' 14 Out of a heroic act to withstand global extractionist agendas, a belief is growing that data can be governed in ways aligned to fantasies of renationalization. 15 The faith in bricks and mortar set in national soil in the form of data centres betrays a wilful disregard of the tendencies of data and digital infrastructures that operate on transcontinental scales through distributed communication networks, at least provisionally so. National dreams of data security and user privacy rub up against the contingency of data and their promiscuous propensity. At the infrastructural level of cables, switch points and Internet exchange hubs, packets of data traffic through clusters of data centres strategically sited according to political economies often tied to state subsidies designed to attract such facilities and the business investment opportunities they promote.

Climatic conditions comprise another key factor that influence where data centres are geographically located and bear upon the operational function and financial cost of running such infrastructures critical to digital economies and society. The cooling systems required to maintain computational systems that transmit data with increasing speed results in what Nicole Starosielski calls a 'regime of thermopower, where the entanglement of media and temperature is shaped by social practices and cultural values'. 16 As the social demand and expectation for data streaming and cloud-based services expands, so too the burden on climate and infrastructure to keep its cool increases in tandem with pressures on electricity generation and energy management.

Data Politics and Sovereign Media

Whether we can organize in political ways as social networks independent of digital platforms and infrastructures is a vexed question. The 2013 Snowden revelations of the National Security Agency surveillance apparatus was a clear prompt for many to review their digital communication practices. Encryption is relatively easy to set up for email, but it becomes more of a challenge for many-to-many modes of communication. One approach adopted by activists involved in Hong Kong's umbrella movement was to engage communication platforms that work off-the-grid. The Firechat app made possible mesh-networks that weren't dependent on activists being online. As localized area networks directly connecting devices, communication and political organization instead happened over Bluetooth connections, which meant that data and information could be exchanged through decentralized networks without being subject to pattern inspection by the enemy due to the database empires in the form of storage facilities such as data centres. Though as an open, unsecured communication protocol, the Bluetooth model did of course leave activists exposed to anyone within range dropping in as infiltrators. Such decentralized practices also raise the question of collective archives as a technology for future political organization. Do activists need to explore decentralization further, building something like small data centres accompanied by the social production of a data commons? For Evgeny Morozov, such a pursuit would make for a radical alternative and legal-social regime to platform hegemony, enabling the 'communal sharing' of data not predicated on extractionist logics and the financialization of daily life. 17

Following the commercial consolidation of the Internet since the late nineties and accumulation and agglomeration of data by tech platforms, decentralized communication practices and architectures are often not an option. And for this reason, along with the multiple hierarchies that comprise any mode of governance – whether they are symbolic, social, architectural, systems-related, or otherwise – it should be enough of a signal for political thought to consider how centralized modes of governance and their attendant infrastructures might serve radical agendas. This doesn't mean jumping on to Facebook to organize your next protest so much as foraging among the infrastructural ruins of digital economy and society and inventing new grammars of expression and polity formation. Perhaps it is time, then, for activists and political theorists of digital media cultures to take seriously the constitutive work of centralized systems of organization, and stop valorizing decentralized, distributed modes of communication. More often than not, these decentred modes are predicated on centralization.

Whether it is high-frequency trading, Bitcoin mining, government administrative and service transactions, or the hosting of social media platforms and their economy of tracking, the question of centralized control remains key to the politics of data infrastructure. Rachel O'Dwyer frames this issue as follows: 'Whoever controls the data centre exercises political and economic control over communications. It's difficult to see how we can counteract these recentralising tendencies in order to build a common core infrastructure'. 18 David Golumbia makes a similar point about high-frequency trading: 'Despite the widespread rhetoric that computerization inherently democratizes, the consequences of the introduction of HFT are widely acknowledged to be new concentrations of wealth and power, opacity rather than transparency of information flows, and structural resistance to democratic oversight and control.' 19 These two observations on data politics are usually separated politically and analytically, but they in fact describe a continuum of common concerns brought into focus through the optic of logistics. Whether it is an anchor to the network of 'flows' or a symptom of the inability of politics to govern the seeming ephemerality of transactional data specific to high-frequency trading, the infrastructural form and technical object of data centres endows data politics with a materiality that otherwise eludes studies of digital culture and society.

We need to talk about the reappropriation of infrastructure and the role of counter-imaginaries in the political geography of data. Can we envisage our own artistic practices or critical research as having some use for colocation services and digital infrastructures? The artistic practice of producing blueprints of an imaginary future provides one technique of subtraction from infrastructural power and the society of tracking, indexing the obsolescence of control and the rerouting of capture into models of psycho-geographic displacement. The case of Hong Kong's umbrella movement demonstrated how infrastructural alternatives work in concrete ways that mobilize populations as political subjects. Tactical media experiments in many settings across Europe and the Middle East further instantiate fleeting utopian desires made real. 20 An especially creative example of collectively imagining futures as performative-pedagogical gestures and diagnostic practices situated in the present, is wonderfully encapsulated in The Future Archive, an ongoing project launched in 2005 by Manuela Zechner. 21 While artistic-activist pursuits such as The Future Archive are often dependent on commercial data infrastructures and content management platforms, such examples suggest that content production is a relevant form of soft infrastructure. In other words, the form of communications infrastructure can accommodate a spectrum of expressions (content, discourse, etc.) While each of these examples is specific to their historical conjunctures, a continuity can also be drawn back to the wilful disobedience of avant-garde experiments and activist pranks of people and groups affiliated with collectives and organizations in the sixties such as the Situationist International, most commonly associated with Guy Debord. 22

Counter-imaginaries are important because they address the question of how to recouple infrastructural design with new forms of collective governance. This is also a question of how to decouple and let go of the fetish of social networking as a form of politics altogether. The delightful field guide of Internet infrastructures of Manhattan, made by Ingrid Burrington, shows how a communication system is registered in multiple forms of semiotic encoding, literally as sidewalk markings, that disclose the location of the Internet's component parts: data centres, different types of underground cables, surveillance cameras, telecommunications companies, manhole covers, cell towers, traffic signal controllers, subway wireless networks. 23 Metahaven's *Black Transparency* is another example of data centre critique, one that combines media theory and political analysis with design aesthetics reminiscent of the incredibly complex and detailed counter-cartographies of political and economic systems by Bureau d'Études. 24

Top: Metahaven, *Transparent Camouflage* (2010–2013), WikiLeaks scarves and T-shirts, installation view at *Frozen Lakes*, Artists Space, New York, 2013.
Photo by Daniel Perez

Bottom: Metahaven, *The Good Ship Anonymous* (2013) and *Large Data Collider* (2013), data centre models, installation view at *Black Transparency*, Future Gallery, Berlin, 2014.

The production of counter-imaginaries are not inconsequential for the politics of infrastructure. Indeed, they are a key component in the invention of concepts that frequently enough migrate as memes from the margins of critical theory and the net-cultural moment of 'sovereign media' to the policy discourses of the state and market strategies of the firm.[25] By Adilkno's account, sovereign media are apparatuses of indifference. They are a negative media of subtraction. 'Unlike the antimedia, which are based on a radical critique of capitalist (art) production, sovereign media have alienated themselves from the entire business of politics and the art scene.'[26] Sovereign media are not consciousness-raising machines. They hold no megaphones. Immanent to media of ubiquity, the dull surface of sovereign media are ideal hosts for the practice of anonymity. They involve a game of tinkering with parameters of the given. They operate within formats of familiarity and flourish when systems short-circuit. Sovereign media are primed to exploit the ruins of a logistical future.

My interest here is to think how infrastructure of communication operate as a form of sovereign media, bringing the singularity of the state as a sovereign entity into question. In an essay written with friend and long-term collaborator Geert Lovink, we made the following observation about communications infrastructure such as data centres: 'Until we know more about the technical operations, communication protocols, legal regimes, design principles and social-economic impact of such infrastructure, the capacity of movements to make informed decisions about how to organize in ways that both support and secure their interests and agendas will remain severely circumscribed.'[27] Sovereign media are not a return to the politics of exodus, but a way to scale autonomy beyond tactical media as demonstrated by WikiLeaks among others. Part of such work involves unleashing alternative blueprints, prototypes and test cases for a future that arises out of infrastructural ruins.

What is to be done with all those data centres shut down in the US during the Obama government's drive for greater efficiencies? We could also add abandoned crypto farms to this list. As Bitcoin burns out and Ethereum stagnates, the churn and burn of energies and ecologies for bodies and biology continue as the scar tissue of crypto-mania. Can such infrastructural remains of data capitalism be repurposed as apparatuses of communication and social-environmental relation without destruction? How might they be put to other uses that don't necessarily result in extending one of the core problems their closure sought to address – namely the wasteful consumption of polluting energy? What if they were powered from sustainable energy sources? Actually this is not an especially powerful line of critique so much as an ironic gesture. As one tech report, dating back to 2013 already noted: 'Apple, eBay, Facebook, and Cisco have already made plans to use of [*sic*] the source of renewable energy for their data centers.' [28] What is required, then, is a repurposing of infrastructure beyond the planned obsolescence inherent in solutionism and its ideology of techno-fixes, [29] designing instead a culture of repair worldwide, as pursued by the numerous maker labs and repair cafés.

One obvious task for designing counter-imaginaries is to engage in the multiplication of communication formats. Template cultures have become today's iron cage of reason. They are an unknown default whose genealogy is not without power in placing limits on expression in seemingly invisible ways. Speculation is no longer the work of imagination but rather is consigned to the operation of machines and the default settings of parameters. We need to collectively orchestrate strategies to infiltrate existing systems and manipulate them for other ends from within. This is not about submitting to the state or any other sovereign entity. But rather, such a move consists of identifying prevailing black box systems of control and collectively devising ways to exploit these apparatuses from within.

This text draws on research from an Australian Research Council-funded Discovery Project, 'Data Centres and the Governance of Labour and Territory'. Thanks to Brett Neilson, Geert Lovink, Soenke Zehle, Rolien Hoyng and Liam Magee for discussions across the time of the project and input into this chapter. For documentation of the research, see *Data Farms: Circuits, Labour, Territory*. Online. Available at *www.datafarms.org* (accessed 28 September 2022).

1. See B. Neilson and N. Rossiter, 'Automating Labour and the Spatial Politics of Data Centre Technologies', in: M. Will-Zocholl and C. Roth-Ebner (ed.), *Topologies of Digital Work: How Digitisation and Virtualisation Shape Working Spaces and Places* (Basingstoke: Palgrave Macmillan, 2021), 77–101.
2. Collocation Data Centers. Online. Available at *www.datacentermap.com/datacenters.html* (accessed 28 September 2022).
3. See L. Magee and N. Rossiter, 'Service Orientations: Data, Institutions, Labour', in: I. Kaldrack and M. Leeker (ed.), *There Is No Software, There Are Just Services* (Lüneburg, Germany: Meson Press, 2015), 73–89.
4. See S. Zehle, 'The Autonomy of Gesture: Of Lifestream Logistics and Playful Profanations', *Distinktion: Scandinavian Journal of Social Theory* (2012) vol. 13, no. 3, 341–354.
5. See S. Apicella, *Das Prinzip Amazon: Über den Wandel der Verkaufsarbeit und Streiks im transnationalen Versandhandel* (Hamburg: Rosa Luxemburg Stiftung, 2021). Online. Available at *www.rosalux.de/publikation/id/45005/das-prinzip-amazon* (accessed 28 September 2022). See also J. Boewe and J. Schulten, *The Long Struggle of the Amazon Employees*, 2nd ed. (Brussels: Rosa-Luxemburg Stiftung, 2019). Online. Available at *www.rosalux.eu/en/article/1558.2nd-edition-the-long-struggle-of-the-amazon-employees.html* (accessed 28 September 2022); Transnational Social Strike Platform (ed.), *Strike the Giant: Transnational Organisation Against Amazon* (Fall 2019). Online. Available at *www.transnational-strike.info/2019/11/29/pdf-strike-the-giant-transnational-organization-against-amazon-tss-journal* (accessed 28 September 2022); H. Geissler, *Seasonal Associate*, trans. Katy Derbyshire (Los Angeles: semiotext(e), 2018); and J. Alimahomed-Wilson and E. Reese (ed.), *The Cost of Free Shipping: Amazon in the Global Economy* (London: Pluto Press, 2020).
6. See N. Rossiter, 'Imperial Infrastructures and Asia Beyond Asia: Data Centres, State Formation and the Territoriality of Logistical Media', *Fibreculture Journal* 29 (2017). Online. Available at *twentynine.fibreculturejournal.org/fcj-220-imperial-infrastructures-and-asia-beyond-asia-data-centres-state-formation-and-the-territoriality-of-logistical-media* (accessed 28 September 2022). See also F. Sprenger, 'Between the Ends of a Wire: Electricity, Instantaneity and the World of Telegraphy', in: M. Hampf and S. Müller-Pohl (ed.), *Global Communication Electric: Actors of a Globalizing World* (Frankfurt: Campus, 2013), 355–438; J. Hills, *The Struggle for Control of Global Communication: The Formative Century* (Urbana: University of Illinois Press, 2002) and D. R. Winseck and R.M. Pike, *Communication and Empire: Media, Markets and Globalization, 1860–1930* (Durham, NC: Duke University Press, 2007).
7. J. Denneen and T. Dretler, 'The Financially Sustainable University', *Bain and Company*, 6 June 2012. Online. Available at *www.bain.com/publications/articles/financially-sustainable-university.aspx* (accessed 28 September 2022).
8. Microsoft, for example, 'refer to core customer data as the customer data that is committed to be stored at rest only within your region according to the Privacy & Security Terms'. They further define data according to the following categories: customer data, personal data, professional services data, administrator data and payment data. Online. Available at *www.microsoft.com/en-us/trust-center/privacy/customer-data-definitions* (accessed 28 September 2022).
9. See L. Parks, *Rethinking Media Coverage: Vertical Mediation and the War on Terror* (New York and London: Routledge, 2018).
10. T.-H. Hu, *A Prehistory of the Cloud* (Cambridge, MA: The MIT Press, 2015), 100.
11. Ibid., 103.
12. See also A. Fish and B.L. Garrett, 'Resurrection From Bunkers and Data Centers', *Culture Machine* (2019) vol. 18. Online. Available at *www.culturemachine.net/vol-18-the-nature-of-data-centers/resurrection-from-bunkers* (accessed 28 September 2022).

13. Online. Available at *docs.microsoft.com/en-us/azure/sentinel/overview* (accessed 11 June 2024).
14. P. De Filippi and S. McCarthy, 'Cloud Computing: Centralization and Data Sovereignty', *European Journal for Law and Technology* (2012) vol. 3, no. 2. Online. Available at *www.ejlt.org/index.php/ejlt/article/view/156* (accessed 28 September 2022).
15. See B. Xiang, 'Renationalization?', blog post, Centre on Migration, Policy and Society, University of Oxford, 25 October 2011. Online. Available at *www.compas.ox.ac.uk/2011/renationalization* (accessed 28 September 2022). See also S. Mezzadra and B. Neilson, 'The Geopolitics of Labour', in: L. Weber and C. Tazreiter (ed.), *Handbook of Migration and Global Justice* (Cheltenham: Edward Elgar, 2021), 14–25.
16. N. Starosielski, *Media Hot and Cold* (Durham, NC and London: Duke University Press, 2021), 212.
17. E. Morozov, 'Socialize the Data Centres!', *New Left Review* 91 (January–February 2015), 64–65. While he doesn't mention it in this interview, one source of inspiration and reference here for Morozov is quite likely the short-lived experiment by the Allende government in Chile in the 1970s, where cybernetician Stafford Beer was commissioned to assist the government in designing Project Cybersyn – a computer system for economic management inscribed with socialist values. See E. Medina, *Cybernetic Revolutionaries: Technology and Politics in Allende's Chile* (Cambridge, MA: The MIT Press, 2011).
18. R. O'Dwyer, 'The Revolution Will (Not) Be Decentralised: Blockchains', *P2P Foundation*, 23 March 2015. Online. Available at *blog.p2pfoundation.net/the-revolution-will-not-be-decentralised/2015/03/23* (accessed 28 September 2022).
19. D. Golumbia, 'High-Frequency Trading: Networks of Wealth and the Concentration of Power', *Social Semiotics* (2013) vol. 23, no. 2, 278.
20. See E. Kluitenburg's efforts here in curating and maintaining a 'living archive' on tactical media. Online. Available at *www.tacticalmediafiles.net* (accessed 28 September 2022).
21. The Future Archive. Online. Available at *www.future-archive.org* (accessed 28 September 2022).
22. For a counter-history of the Situationists, see M. Wark, *The Beach Beneath the Street: The Everyday Life and Glorious Times of the Situationist International* (London and New York: Verso, 2011).
23. I. Burrington, *Networks of New York: An Internet Infrastructure Field Guide* (New York: Eyebeam Art and Technology Center and Data and Society Research Institute, 2015). An expanded edition was later published by Melville House in 2016.
24. See Metahaven, *Black Transparency: The Right to Know in the Age of Mass Surveillance* (Berlin: Sternberg Press, 2015) and Bureau d'Études. Online. Available at *www.bureaudetudes.org* (accessed 28 September 2022).
25. On the concept of 'sovereign media', see ADILKNO, *The Media Archive: World Edition* (New York: Autonomedia, 1998), 12–15. See also C. Apprich and F. Stalder (ed.), *Vergessene Zukunft: Radikale Netzkulturen in Europa* (Berlin: transcript Verlag für Kommunikation, Kultur und soziale Praxis, 2012) and C. Apprich, 'Remaking Media Practices: From Tactical Media to Post-Media', in: C. Apprich, J. Berry Slater, A. Iles and O. Lerone Schultz (ed.), *Provocative Alloys: A Post-Media Anthology* (London: Mute Books, 2013), 122–140. Online. Available at *www.metamute.org/shop/mute-books/provocative-alloys-post-media-anthology* (accessed 28 September 2022).
26. ADILKNO, *The Media Archive*, op. cit. (note 25), 13.
27. G. Lovink and N. Rossiter, 'Network Cultures and the Architecture of Decision', in: L. Dencik and O. Leistert (ed.), *Critical Perspectives on Social Media and Protest: Between Control and Emancipation* (New York: Rowman & Littlefield, 2015), 227.
28. P. Jones, 'US Data Center Turns to Solar for Power', DatacenterDynamics: The Business of Data Centers, 29 August 2013. Online. Available at *www.datacenterdynamics.com/app-cloud/us-data-center-turns-to-solar-for-power/81850.fullarticle* (accessed 28 September 2022).
29. See E. Morozov, *To Save Everything, Click Here: The Folly of Technological Solutionism* (London: Allen Lane, 2013).

Hyperscaling the Public Interest: How the Public Stack Can Help Put Society at the Core of Large-Scale Tech Development

Max Kortlander, Marleen Stikker
and Sander van der Waal

Secret Plans for a Hyperscale Data Centre in Zeewolde

A remarkable debate started in the Netherlands near the end of 2021, when plans to build a hyperscale data centre in the small municipality of Zeewolde were made public. The ruckus began with the forced publication of hidden documents through a Freedom of Information request by journalists, and was fuelled by the fact that the plans involved one of the most contentious technology companies out there: Facebook, or Meta/Facebook. [1] Representatives from Meta/Facebook and local officials had secretly agreed to develop a hyperscale data centre on a large tract of farmland, spurring the local council to rezone the agricultural land as an industrial area. [2] The plan's revelation sent a shockwave through the small town that reverberated across the country. What had previously been a calm agricultural area was now subject to radically change, all without citizen knowledge or input. Their environment was set to transform into a closed technological space to be exclusively used by Meta/Facebook.

Until then, few people paid attention to hyperscale data centres – such matters were only discussed by an inspiring coalition of artists, hackers, researchers, journalists, environmentalists and activists – but this was a breaking point. The resulting public debate over the Meta/Facebook hyperscale data centre took place on talk shows, in intense Twitter feeds, through social media campaigns and via statements from national politicians.

Questions about every aspect of data centres were now on the table: the electricity consumption (equivalent to 460,000 households); the allocation of land to be utilized by a single Big Tech company (what about farmers or the need to build houses?); the secrecy of the process (years of preparation under a code name); the political eagerness to make it happen ('We will play in the Champions League' as one alderman proudly said); the rationale for prioritizing Meta/Facebook in the transition to green energy, which is still scarce (the data centre would consume an amount of green energy worth 10 per cent of current wind energy capacity); and the biggest question of all – how could 20 part-time local politicians make a decision of national importance without a public process?

The process to develop a Meta/Facebook hyperscale data centre in the Netherlands did not have to be closed until forced open. We demand much more in our democratic society. The Fourth Rutte cabinet was taking control to decide on a national level where hyperscale data centres can be located, and decided that hyperscales can only be allowed in three areas in the Netherlands: Zeewolde is not one of these areas, meaning that the plans to build a hyperscale data centre there have mostly been scrapped. However, there is still a risk that these national level decisions will continue to prioritize narrow economic considerations over societal concerns. In order to put the public interest at the core of these deliberations, we need to revisit our foundational assumptions, revise our approach to design and technology, and reframe the role of people and society in relation to our shared digital trajectory.

What Is the Role of Data Centres?

Data centres are absolutely necessary in order for crucial aspects of the Internet to function. They store large amounts of data, providing redundancy and backup of data and digital systems; they facilitate the movement of data by providing infrastructure for virtual networks; and they provide the computing capacity to run the systems and algorithms necessary for hosting critical digital tools and services. Operating at an economy of scale, data centres pool resources to enable the technological processes that we expect, need and depend upon in business, government and our daily lives.

Not all data centres are equal, however. Instead, we can and should judge a data centre by the extent to which it serves shared societal interests rather than private or corporate interests. A starting point in this regard is to consider how power and control over essential technical processes correlate with power and control more generally in society. In order to open or democratize the technical processes around data centres, it is necessary to question their foundational purposes as well as their effect upon people and the planet.

We should consider *which* actors are involved, and *how* they are involved – whether as decision makers, active users or passive data sources. We can also consider people's agency in relation to the effects of a data centre: do they know what data is stored about them, and how it is used? And of course, we must consider who benefits and who does not.

When approaching such questions, the relationship between a data centre and its direct environment is unavoidable. There are of course local issues related to laws, for example, about how personal data is gathered and used. Furthermore, data centres occupy a large physical presence, and have a corresponding ecological impact in terms of land, water and electricity consumption. The buildings themselves are often closed to the public, and such exclusivity poses inherent problems for democratic societies by restricting the extent to which the public can judge or otherwise engage with them.

There is a crucial difference between data centres that are open to be used by a variety of actors and so-called 'hyperscale' data centres. Hyperscale is perhaps not the right word to capture the nuance between different types of data centres, because scale is not the issue at hand – there could presumably be a data centre that is large in scale and also serves the public interest. Instead, the issue with hyperscale data centres is rather that they tend to be built to serve the needs of a single company, as is the case with the proposed data centre in Zeewolde. The function of such data centres is particularly problematic when based upon a 'surveillance capitalist' model, whereby a company or individual profits at the expense of privacy and autonomy.

If a data centre is not built and operated in accordance with the public interest, then something needs to change. Considerations about the extent to which a data centre serves the public interest can become complex when there are multiple layers at play – considerations about values, design processes, technological processes and the impact that all of this has on people and the planet. The public stack model is one lens through which we can approach the complexity of such layered considerations, in order to further illuminate the rights and responsibilities in the interaction between a data centre and society.

The Public Stack

We can think of technology in terms of stacks. At the top of that stack is you, the citizen, interacting with a website, a phone or a surveillance camera in your city. But this top layer is only the tip of the proverbial iceberg. Beneath it, and often unseen, are a number of factors that determine how a person or community relates to that technology. Firstly, there are other technology layers that you hardly ever see: the various components inside your smartphone or laptop, the Wi-Fi routers that enable connectivity, and the marine cables on the bottom of the ocean that connect all of the continents physically. Secondly, all of these technologies have been designed, developed, built and constructed based on defined design and development processes. Finally, there are foundational issues determining the basis for development: these include the starting points and assumptions on which the technology has been developed; how human rights and public values are respected in it; how governance and supervision have been organized; and what socioeconomic model is underpinning the technology. All of these factors determine the position of people in relation to technology and can be considered together as a stack comprised of four main layers: the *foundation*, *design process*, *technology* and the *citizen* layer.

There are different kinds of stacks. A lot of the technology we encounter – your Instagram account, the microphone in your rental car, your neighbour's new Ring Video Doorbell – is the result of a *private* stack.[3] The foundation of these technologies (their *raison d'être*) is for profit and built upon profit-motivated assumptions. Everything about them is proprietary. They are the result of closed design processes and black-box technology, based on commercial market values. At best, the people encountering these technologies are 'consumers', and all too often, they are 'consumed'.

You may also be familiar with technology that is the result of a *state* stack: think of the series of scanners you go through at an airport (not just the physical scanners you put your body into, but also the entire invisible layer that scans your identity and travel history). Like the private stack, technology stemming from the state stack is similarly closed and optimized for surveillance, ultimately positioning people as 'subjects'. The state stack and private stack are deeply intertwined, like when a government collects data gathered through private platforms, or in the way that governments themselves are often 'locked-in' and dependent upon private vendors.

There is a need for an alternative to private and state stack technologies – not just alternative software and hardware, but an alternative framework for conceiving, designing, building, delivering and using technology – and this is where the public stack comes in.

In the public stack, the foundation is based on common values, shared goals and publicly deliberated assumptions of a city, community or any group with shared concerns. The foundation guides and supports how design choices are made, how the technology is designed and documented, and how all of this positions *people* or *citizens* who use, interact with, or are otherwise affected by

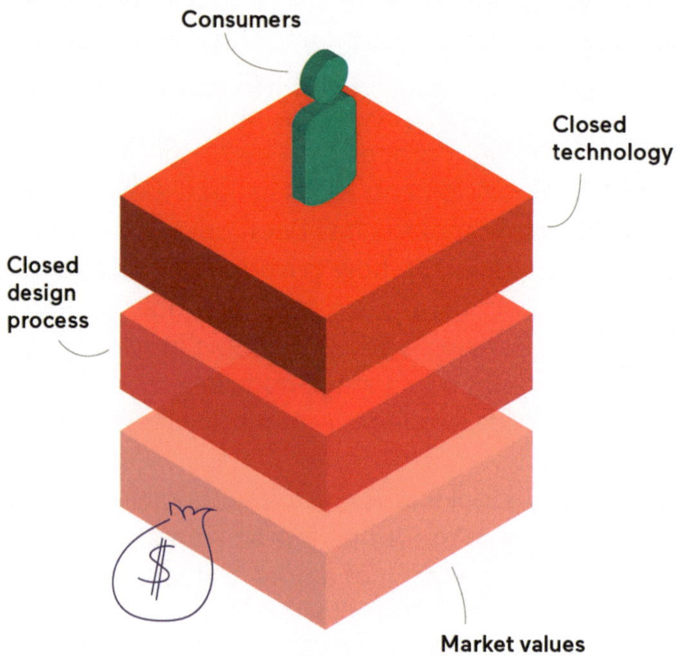

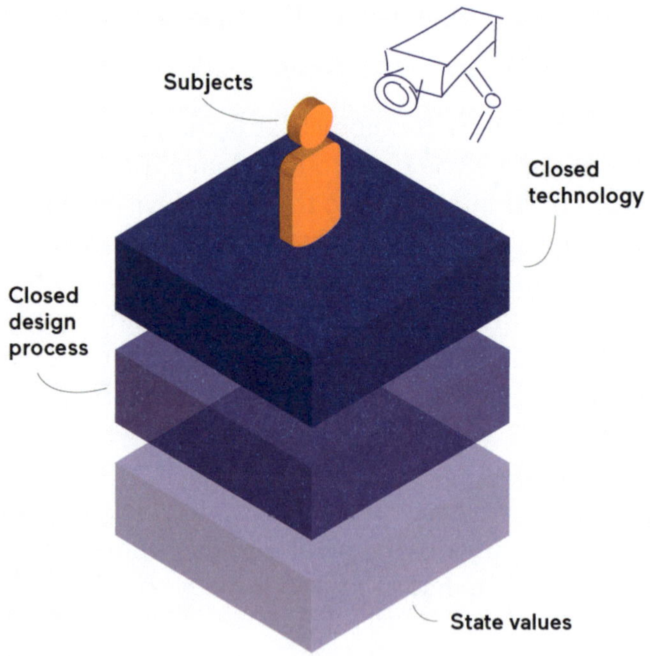

The foundation of the private stack (top) is profit-driven and based on market values. The foundation of the state stack (bottom) is based on centralized state values. Surveillance is central to both models.

the technology. When the public stack is applied within a society that shares democratic values, it follows that subsequent design processes ought to be open and participatory for citizens through co-creative methods. The technology itself must be open source, not a black box: how else could people democratically judge that technology and its influence? Ultimately, the various layers of the public stack ought to culminate in positioning people as free in relation to both technology and society.

Assumptions: The Private Stack in Practice

The case in Zeewolde is an example of the private stack in practice – a process that was optimized for private equity at the cost of public values. The situation at hand, however, merits a public stack approach. A first step in this regard is to fully disclose the underlying assumptions, myths and crafty narratives that legitimize choices made (by the few) about technology. In other words, we need to begin our transition to a public stack approach by revisiting the *foundations* of our current tech development.

Assumption: Data Is (Always) Good

There is a general assumption that our future is digital: that data and algorithms are necessary to solve the problems of humanity, and that more data is always a good thing regardless of who is collecting it, who it is hidden from or how it is used. It is said that technology will make our society smarter, safer and more sustainable; that we will be able to foresee and control our future with data-driven decision making; that data is the new gold, the new oil. From these myths, it follows that data centres are an expression of smart innovative regions. Big Tech companies are essential, so we should be proud to host them on our soil.

Within this frame of mind, closed and proprietary hyperscale data centres are inevitable. Regions and countries thus compete for the data centres, as politicians believe they act in the general interest when the subsidize, prioritize and remove obstacles for tech companies.

Data is not good in and of itself. People assume digitalization is good for society, but we have to be nuanced: there are many ways that data can help us to understand the world better, but data and algorithms are *always* expressions of power. We have to carefully choose what we define as valuable data – data that we want to gather and analyse. Currently, choices about this are not ours. Priorities and decisions about what data to collect, who collects it, and how it is used are made behind closed doors and without input from society. We thus tend to invest in technology that accumulates the wealth and power of Big Tech companies at society's expense.

Assumption: Digital Is Virtual

People tend to use 'digital' and 'virtual' synonymously, as if digital technology is somehow not real, not part of our real, physical world. We thus treat issues involving digital matters differently than we treat more tangible matters. The problem with this is that digital technology *is* real, and it *is* embedded and growing in our real, physical world – to think otherwise is a huge mistake.

Digital technology is tangible, despite the illusion of its intangibility. There is an enormous cost in building digital tech and infrastructure, which also tends to perpetuate and exacerbate conflict and inequality through unfair labour and trade conditions as demonstrated, for example, by moderation models wherein moderators in developing countries 'must watch videos of murders, rapes, suicides, and child sexual abuse'.[4] Scarce natural resources live short lives in technology that quickly becomes outdated, obsolete and ultimately 'e-waste'. The electricity consumed by information and communications technology is projected to account for one fifth of global energy consumption by 2030.[5]

Digital technology is also a part of our physical world in terms of data flows and surveillance, and thus is a very real part of our society and culture. The global surveillance network is embedded in our 'smart' city centres, in our 'smart' cars, in our 'smart' homes, and even in the 'smart' phones in our pockets. It is present in nearly every part of our physical worlds, from our shared public spaces to our personal private spaces, making public spaces much less public and private spaces much less private.

Hyperscale data centres take up space, use energy and form a significant part of the private and state surveillance structures that monitor us throughout our homes and cities. Like other technology, data centres are physical, real and should be treated as such.

Assumption: Increased Economic Activity Is Always in the Public Interest

Governments tend to hold the pervasive assumption that they should accommodate the economic activity and growth of private parties because, at some point in time, this will provide value for society. However, having a hyperscale data centre on a Dutch polder does very little for our national economy let alone for our society. In many cases, economic decision processes measure success in terms like jobs added, increased real estate prices and cheaper services.

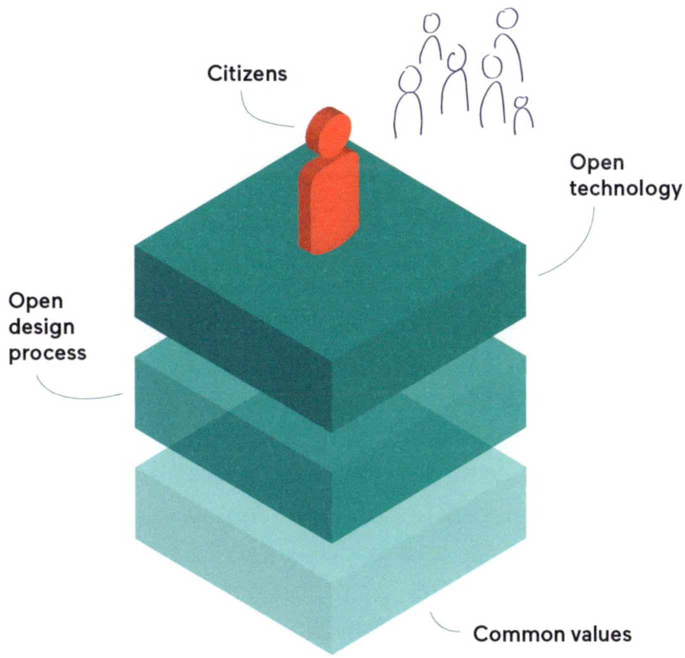

The public stack is based on common values, resulting in an open design process and open technology.

It is a disproven economic theory,[6] but liberals, conservatives and social democrats alike maintain its assumptions as an underlying reason for policy decisions.

We support statements by economists such as Kate Raworth who argue that all economic activity should be regenerative, because we live on a finite planet.[7] Extractive business models should be replaced by those which promote health, peace, well-being, equality and environmental sustainability. A Meta/Facebook hyperscale data centre is incompatible with a responsible economy.

Applying the Public Stack to Hyperscale Data Centres

The assumptions we hold about data, technology and economics are all at play, but are they true and are they beneficial? What would happen if we abandon current myths and assumptions, and adopt a public stack approach that bases its foundation in shared values, common goals and publicly deliberated assumptions? How should the subsequent design processes, technologies and positions of people and society be different?

Imagine that the *foundation* for major tech developments (like the Meta/Facebook data centre) would include shared democratic values that put society in control over its own future. This would require us to rethink our assumptions about what we optimize through data and economics, and to reframe which stakeholders and resources merit inclusion. When we consider who a data centre's stakeholders are, we do not just need to include the people that live in the region where the data centre would be built, but may also need to include everyone in the country if its development and operation affects national matters. When we consider whose data is stored, Meta/Facebook wanted the data centre to service all of Europe and the MENA-region. The resources at stake merit inclusion of considerations that go far beyond those that are profit-based, and we thus must rightfully consider so-called 'externalities' like working and social conditions, privacy and human rights, and the environment.

Proper inclusion of relevant stakeholders and resources requires a comprehensibly inclusive *design process* that is co-creative, participatory and open to all people. We need to approach the proposition of a hyperscale data centre and other large-scale tech development by bringing together a variety of competencies that open decision-making processes to public scrutiny and collaboration. As with these other design process, we should arrive at an integrated vision that defines what we allow within the boundaries of people and planet:[8] how many data centres we allow, with which requirements and under which conditions.

An open design process also requires fully transparent use of *technology*. This means no proprietary knowledge, hidden documents or closed secret discussions. There is not a single opening in the way Meta/Facebook operates: all of their technology and infrastructure is part of one big black box and actively works against the public values by extracting data from people, developing profiles about their preferences and fears, and deliberately manipulating them and thereby influencing our public debate in ways that we cannot investigate. 9 This type of behaviour is not inherent to data centres: there are good reasons to have them, and good ways to develop them, like Greenhost, which explains clearly how data is handled on their servers. 10

Citizens should not be exploited for their data. The planet should not be exploited for private gain. These are general starting points for all technical and urban development plans that have been largely ignored by developers of hyperscale data centres.

Most data centres are privately owned and operated. For uses like government services, there are indeed already public data centres. One potential way forward to is expand public data centres to include openness for use other than governmental purposes. This line of thought raises important questions about sovereignty and control over public digital infrastructures: do we expect governments to create open public data centres, or do we envision other ways to develop open, fair and inclusive public digital infrastructures (data centres included)?

Whichever route we choose, our decisions about where to place data centres and how to build them need to take better consideration of the local and global environments. We ought to prioritize building data centres in areas that minimize negative environmental impact, for example, by placing them in areas with natural cooling or natural sources of energy. Research efforts like Microsoft's Project Natick, 11 whereby a data centre was placed underwater to take advantage of natural cooling, should continue to be pursued as we refine our understanding of how to build and maintain more sustainable data centres.

Conclusion

We will undoubtedly continue to face decisions about not only large-scale tech developments, but also about other issues related to energy, water and the environment. Decisions made in these areas – including decisions about hyperscale technology – affect all of us. Our shared living environment, shared energy system and shared values are all at stake in developments like Meta/Facebook's hyperscale data centre.

For the services we want and need, there is no feasible solution other than large-scale data centres. Our policies in developing them thus far, however, fall short of properly accounting for their impact on people and the planet. Fortunately, the policy approach towards hyperscale data centres and other matters of shared importance is not yet set in stone. That means it's not too late for us to reframe policymaking processes around technology – as well as other overarching matters of environment, security and human rights – to rightfully include citizens as an integral part of democratic societies. The political process around our future digital and environmental trajectories should have been public and open to everyone all along: let's make it that way now.

1. Officially the company has changed its name to Meta, but we refuse to play along with this crafty PR move!
2. See T. Brown Hamilton, 'In a Dutch Town, a Fight With Meta Over a Massive Data Center', *Washington Post*, 1 June 2022. Online. Available at www.washingtonpost.com/climate-environment/2022/05/28/meta-data-center-zeewolde-netherlands (accessed 5 July 2023).
3. 'Encounter' and not 'use' because we often interact with technology without giving consent or making a conscious choice to do so.
4. B. Perrigo, 'Inside Facebook's African Sweatshop', *Time*, 14 February 2022. Online. Available at www.time.com/6147458/facebook-africa-content-moderation-employee-treatment (accessed 5 July 2023).
5. N. Jones, 'How to Stop Data Centres From Gobbling up the World's Electricity', *Nature*, 13 September 2018. Online. Available at www.nature.com/articles/d41586-018-06610-y (accessed 5 July 2023).
6. J. E. Stiglitz, 'Inequality and Economic Growth', *Rethinking Capitalism* (London: Wiley-Blackwell, 2016), 134–155.
7. See *Kate Raworth: Exploring Doughnut Economics*. Online. Available at www.kateraworth.com/doughnut (accessed 5 July 2023).
8. Ibid.
9. Facebook is notorious for not sharing data about the workings of their platform with researchers, and shutting down researchers who attempt this. See N. Kayser-Brill, 'Algorithmwatch Forced to Shut down Instagram Monitoring Project After Threats From Facebook', AlgorithmWatch. Online. Available at www.algorithmwatch.org/en/instagram-research-shut-down-by-facebook (accessed 5 July 2023).
10. Greenhost's privacy policy prioritizes 'privacy by design and complies with European data protection legislation'. Purposes of data collection and information regarding erasure of data are clearly stated. See 'Privacy Policy', Greenhost. Online. Available at www.greenhost.net/legal/privacy-policy (accessed 11 July 2023).
11. J. Roach, 'Microsoft Finds Underwater Datacenters Are Reliable, Practical and Use Energy Sustainably', Microsoft, 14 September 2020. Online. Available at news.microsoft.com/innovation-stories/project-natick-underwater-datacenter (accessed 5 July 2023).

Materialities – The Physical Manifestations of the Cloud

We are becoming increasingly aware that digital culture, or digital capitalism, is far from intangible. The technology, data centres not least, from which the digital emerges is material and partly very earthly. It generates significant amounts of e-waste, consumes large amounts of energy, and contributes to climate change. 'Materials' focusses on the physical manifestations and environmental footprints of cloud computing and data centres. Hence, the key questions in this chapter are: what are the physical and environmental impacts of data storage and its structures on our living environments, and what inequalities do they perpetuate?

As an information designer and co-author of the nighttime photo series *Acid Clouds*, **Niels Schrader** seeks to understand what happens when the resources needed to create, process, distribute and store our daily output of 2.5 quintillion bytes of, we assume, 'virtual' data encroach on the physical environment. [1] In his research, he pays particular attention to Amsterdam's role as a major data centre hub and the fact that the IT sector is a predominantly white, heterosexual, male industry. In search of a more balanced information ecology, Schrader questions the growth narrative of the tech industry and appeals to our responsibility as users.

Artist and academic **Füsun Türetken** explores the agency of matter, in particular the role of metal in shaping the world of capitalist finance, belief systems, geopolitical relations and (digital) bodies. Through a history of 'the Soviet Internet that never came to be' and the role in it of Alexander Brodsky and Ilya Utkin, members of the Soviet artists' collective Paper Architects, she argues that capitalism leaves behind a planet suffering from 'Gaiaporosis', eroded by extractive economies.

1. B. Marr, 'How Much Data Do We Create Every Day? The Mind-Blowing Stats Everyone Should Read', *Forbes*, 21 May 2018. Online. Available at *www.forbes.com/sites/bernardmarr/2018/05/21/how-much-data-do-we-create-every-day-the-mind-blowing-stats-everyone-should-read* (accessed 29 July 2024).

Acid Clouds

Niels Schrader [1]

Storage as a Concept

The number of computer data storage facilities has grown exponentially in the last few years. This is no surprise given the global increase in consumption of digital content, propelled by seemingly unlimited bandwidth and ubiquitous Wi-Fi coverage. However, little attention is being paid to the ecological effects of these endlessly humming, heat-producing and energy-draining architectural structures now often among the biggest CO_2 polluters of the environment. [2]

Server farms are large-scale repositories of data that require a massive maintenance framework. The agricultural reference in the name is no coincidence. With the rise of capitalism, farmers began to produce for the purpose of profit accumulation instead of personal or local consumption, and crops became a commodity to be traded. In the supply chain, storage has always played a key role. It made speculation possible, allowing for the sale of agricultural products in times of scarcity and their purchase during a market surplus. Simply put, it became possible to create an artificial shortage that leveraged the demand. Data centres are the new control rooms of the automated landscape, which are not only storing the growing virtual knowledge of the world, but also facilitating its distribution and capital flow. Behind high, windowless walls, the seemingly ephemeral, clean image of data solidifies into a much heavier and hardly degradable, resource-intensive condition, with growing implications for society.

Growth Narrative

In the technology sector, the capitalistic promise of infinite economic growth is locked tight to Gordon Moore's observation that the number of transistors on a chip doubles every year while its costs are halved. 3 It describes the rapid growth in computational power over decades. Similar dynamics emerge in the global demand for data storage and it seems that both are interlinked. The faster you can process data, the bigger your appetite for it. Conversely, the more data becomes available, the more computational power is required for storage and processing. The industry creates incentives for users not only to produce content, but also to share and back it up, meaning in fact to endlessly duplicate and never delete. Continuous storage is labelled by the tech industry as the solution to our storage problem, but it is in fact creating it instead. The more kilo, mega, giga, tera, peta or exa, the more we unlearn to control our digital clutter. Clutter that manifests itself as overly full inboxes with unanswered emails, deserted social media accounts we never actually visit anymore, long-forgotten download files clogging disk space and abandoned Dropbox folders storing identical data on multiple devices.

'Data is the pollution problem of the information age', says Bruce Schneier, cryptographer and computer security specialist, and fellow at the Berkman Klein Center for Internet & Society at Harvard Law School, Cambridge, Massachusetts. 4 Digitalization creates an ever-increasing amount of data that society will eventually need to learn to dispose of. So even though IT efficiency is improving and data centres are becoming greener and consuming less power, these energy savings are offset by growing electricity consumption.

Physical Data

The IT sector labels its products using metaphors that suggest their association with the natural world and the qualities of cleanliness and transparency, thus covering up or deflecting attention away from the damaging effects of technology on our living environment. Probably the most familiar of these deceptions is the projected image of a fluffy, vaporous 'cloud' used to misrepresent the power-consuming and earthbound qualities of digital data storage. Similarly, tech companies use visual and verbal rhetoric to present their large-scale storage facilities as clean, sterile and environmentally friendly domains. Google, for example, publishes on the web over-styled press images of its data centres in order to shape public perception.

There is a similar dissonance between the image of how tech companies market their services and the reality of what happens to consumer data. The sharing

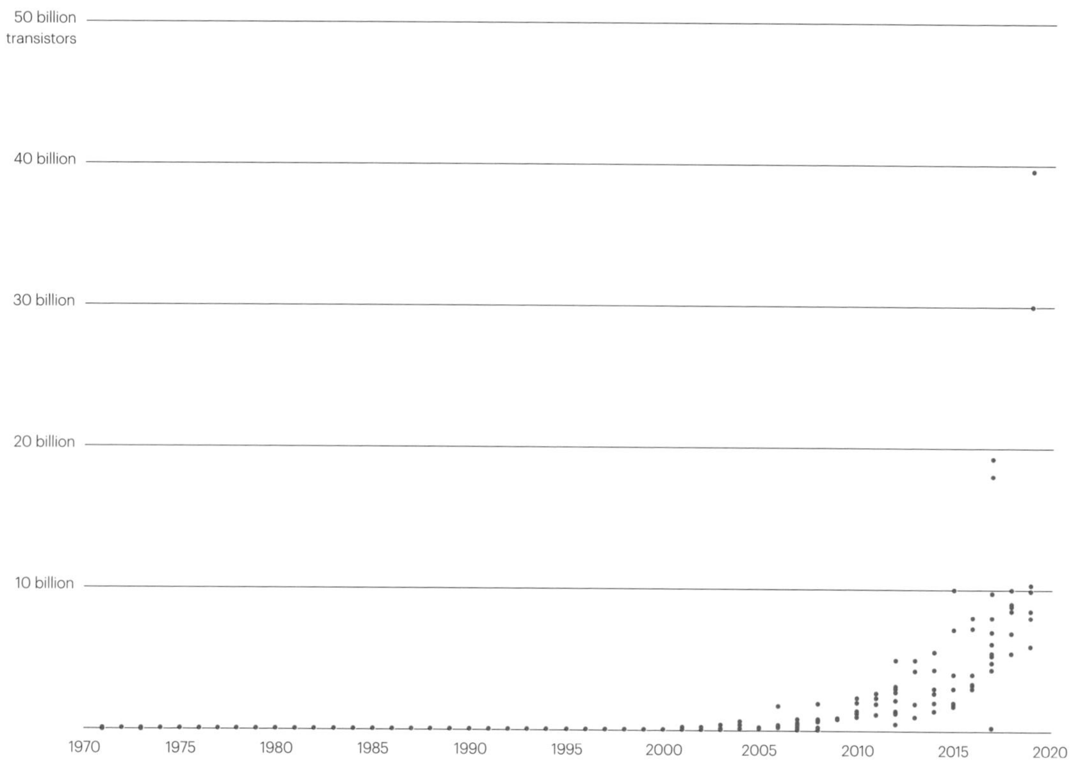

Moore's Law – Number of transistors on integrated circuit chips.
Source: Wikipedia

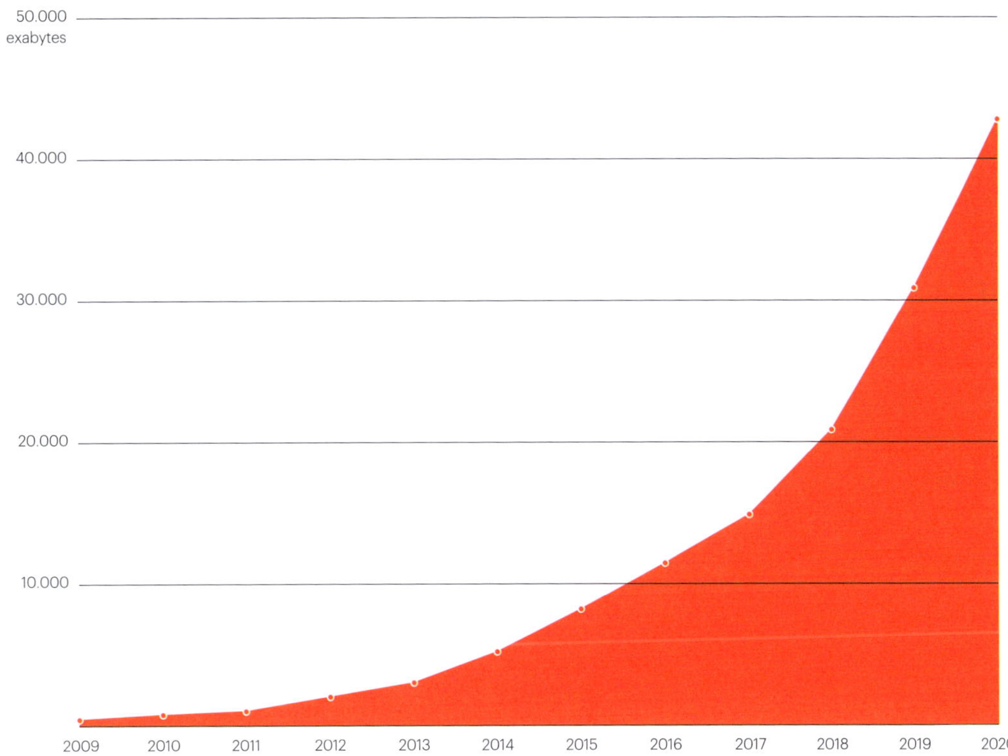

World data storage demand in exabytes
from 2010 to 2020.
Source: Statista

economy, constant online access and subscription-based models are presented as virtues, having obvious benefits for users. But in reality, such qualities provide a frontend distracting from the real activity of data harvesting and offloading both storage and ownership to software companies. Having data is key and keeping it even more so. To ensure that, server farms are highly protected spaces with concrete demarcations and tangible security measures. The buildings rarely have windows and are commonly surrounded by electric fences, security cameras and water moats. On the inside they are packed with the latest security gadgets like motion detectors, iris scans, facial recognition systems or personalized access codes. It seems the physical access to the data is as regulated, monitored and restricted as the virtual one.

Today's digital landscape is characterized by vast stores of loose, unrefined data, heavily harvested by national intelligence services and large private corporations to track people's movements and behaviour. The typical business approach taken by these corporations is to capitalize on the fact that storage is cheap and to keep on accumulating data even if they haven't yet determined a specific use for it.

Information Ecology

The environmental impact of data centres occurs in three main ways: electronic waste, energy consumption and carbon emissions. It is surprising how tangible the effects of these are.

The term e-waste covers the physical hardware inside the building and its recycling or disposal. According to *The Global E-waste Monitor 2024* published by the United Nations University, the global amount was expected to grow to 52.2 billion kilograms in 2021, with an annual growth rate of 3 to 4 per cent.[5] This number was even surpassed in 2022 by a record of 62 billion kilograms.[6] Propelled by short, market-driven replacement cycles, the amount of obsolete equipment is steadily increasing.

The energy consumption on the other hand concerns the power that is required to run the digital framework of the facility. That includes powering the actual computers, but also the cooling infrastructure. The *Independent* reported that in 2015 the world's data centres consumed about 3 per cent of the global electricity supply – a costly problem, which in 2008 translated to roughly an annual cumulative amount of $4.5 billion for data centre-related energy bills for the United States alone.[7] The third environmental impact of the data centres are the CO_2 emissions. Clearly, they are not directly produced by the data centre itself, but by its energy supplier. If a data centre runs on fossil fuels, the equivalent of a new small city is added to the power grid every time somebody opens one of these facilities. Data centres accounted in 2015 for 2 per cent of worldwide greenhouse gas emissions. This is about the same carbon footprint as the airline industry.[8]

It becomes clear that even though data is virtual by nature, we will reach the point where digital concerns hit physical constraints. Continuing along this path will leave a giant ecological and economic footprint. The Cisco Global Cloud Index predicted data centre traffic to jump to 15.3 yottabytes in 2020, a figure with 24 zeroes.[9] As a result, the US Energy Information Administration expected global energy consumption to grow by about 53 per cent by 2035.[10] A concrete example of how tangible the impact of data centres on our living habitat has become, is Amsterdam's acute capacity problem with the local power infrastructure.[11] The city's ambition to become Europe's biggest data centre hub and its reputation as a start-up hotspot are starting to take their toll.[12] As for Amsterdam claiming to be the number one digital gateway to Europe, its regional power grid is just not growing fast enough to keep up with all the new technological developments. As a result, the reserve capacity is used to stabilize the electricity supply and putting other municipal operations under pressure. It remains to be seen how further growth of this industry will be brought under control.

Solutions for this challenge are highly unlikely to come from either the big businesses that directly profit from our excessive behaviour or from the states and governments that show just as keen an interest in our personal data. We should not fall into the trap of believing that increasing the degree of automation and making energy sources greener will solve the problem. Yet, blaming the storage industry solely would mean deliberately ignoring the responsibility that we as consumers have. Even the most eco-friendly technologies and highest recycling rates cannot compete with the simplest but most effective strategy: reducing the production of data and avoiding digital waste. It is about time to replace the growth narrative, acknowledging the fact that we *are* in fact part of the problem and learning how to control and dispose of our digital clutter.

Postscript

After many years of research, it seems that the only way to tackle the challenges related to the Dutch digital landscape will be to re-establish a bare minimum level of transparency and trust in the triangle of stakeholders: the government, the industry and the public. The protests of local activists against Meta's plan to build a massive hyperscale data centre near Zeewolde,[13] and the hot-blooded dispute between Microsoft and farmers from the municipality of Hollands Kroon,[14] make it painfully clear that public citizens need to get a seat at the table of actual decision-making. Also, governmental institutions need to take their

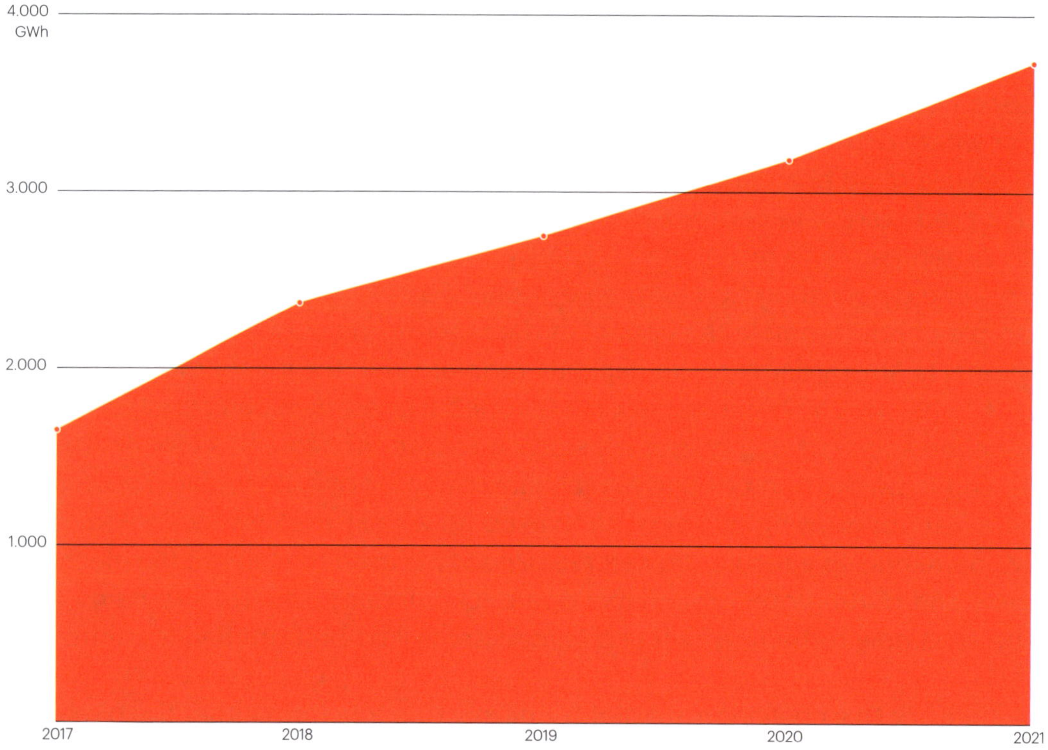

Data centre energy consumption
in the Netherlands from 2017 to 2021.
Source: Statistics Netherlands

digital duties more seriously. Dismissing the need for clear digital legislation has real-life impact on societies. In 2020, for example, administrative negligence became chillingly obvious in a parliamentary inquiry during which former Minister of Economic Affairs and Climate Policy, Eric Wiebes (VVD), brushed away a request for clarification by responding that it is not the national government who is responsible for the supervision of zoning and energy consumption of data centres, but the local authorities. 'This is ultimately a local decision and I will respect that', was Wiebes's response to the motion by Member of Parliament Sandra Beckerman (SP). 15

The tech industry, on the other hand, will need to change their modus operandi, too. Their blind belief in capitalist profit schemes converts every computational operation into a commodity and locks users into an endless cycle of digital dependencies. The streaming, the storing, the sharing, the shopping, the gaming, these are not the services consumers necessarily asked for but the ones they are lured into. By embodying a compulsive believe in innovation, each time Silicon Valley dumps one of their highly anticipated novelties back into the market, a new demand is created. As long as the tech companies continue to believe in the limitlessness of digital expansion, they will continue to undermine regulations of public authorities and will ensure that customers not have a say. It's a cash cow to be milked at every opportunity. Techno-optimistic views that only technology can solve the problems caused by technology is nothing more than a business model. What in fact needs to happen, is a radical shift towards public participation, comparable to the open-source movement in computer programming.

The users on the other side, need to realize that they are complicit with the extractive logics of the data economy. Under the banner of self-realization and -optimization, consumers follow the call of the digital sirens way too easily. They purchase, they consume, they discard and the cycle starts over again. Once society acknowledges its own culpability, scaling down seems to be the most obvious step to take towards a just and green future. From a consumer's point of view, Moore's Law has led to an exponential growth in technology and, consequently, to a monumental increase in data. The digital acceleration induced by machine power now threatens to outrun human-level intelligence, not because it's smarter, but because it's faster, and more efficient. The speed and sheer volume of data transfer is surpassing the level of human understanding and exceeding the controllable threshold. The only possible response to that is to slow down and make available the time for checks and balances. Unplugging for a healthier environment, but also for a healthier lifestyle.

Public authorities will also benefit from a more expanded bottom-up approach to digital sovereignty. It will create more participation when addressing challenges related to the energy and housing crisis, and create more oversight on the money flow. It will help prevent corruption like in the case of Dutch real estate entrepreneur Rudy Stroink and also include local farmers and affected communities in the discussion with Big Tech. [16] Rudy Stroink and his wife were sentenced to community service in 2018 for bribing an American Google director. After a legally imposed nine-month moratorium, since December 2023 a long-awaited legislative amendment prohibits the construction of hyperscale data centres on all Dutch soil. However, this provision excludes the region Hollands Kroon in the province of Noord-Holland and Het Hogeland in the province of Groningen, the top data centre hotspots in the Netherlands. [17] It seems that despite the early adoption, governments keep playing catch-up. No matter which perspective one takes on the mesh of opinions and concerns related to data centres, a meaningful level of participation and transparent information flow is the only way to gain basic credibility. Ultimately, it was also the Dutch government's favourable tax climate that convinced hyperscale businesses to settle in the Dutch polders.

1. This is an edited version of N. Schrader, 'Acid Clouds', *Open! Platform for Art, Culture & the Public Domain*, 11 September 2019. Online. Available at www.onlineopen.org/acid-clouds (accessed 19 April 2024).
2. See R. Koops and B. van Zoelen, 'Dit zijn de tien grootste CO_2-vervuilers in Amsterdam', *Het Parool*, 27 May 2019. Online. Available at www.parool.nl/ps/dit-zijn-de-tien-grootste-co2-vervuilers-in-amsterdam~b3e624a1 (accessed 10 April 2024).
3. See G.E. Moore, 'Cramming More Components Onto Integrated Circuits', *Electronics* (1965) vol.38, no.8.
4. See B. Schneier, panel discussion, *Guardian Activate*, New York, 28 April 2011. Online. Available at www.theguardian.com/activate/video/activate-new-york-bruce-schneier (accessed 10 April 2024).
5. See C.P. Baldé et al., 'Global E-waste Status and Trends', *The Global E-waste Monitor 2017* (Bonn/Geneva/Vienna: United Nations University/International Telecommunication Union/International Solid Waste Association, 2017). Online. Available at www.collections.unu.edu/eserv/UNU:6341/Global-E-waste_Monitor_2017__electronic_single_pages_.pdf (accessed 10 April 2024).
6. See Baldé et al., *Global E-waste Monitor 2024* (United Nations Institute for Training and Research, 2024). Online. Available at api.globalewaste.org/publications/file/297/Global-E-waste-Monitor-2024.pdf (accessed 29 July 2024).
7. See P.W. Taylor, 'How to Cut Data Center Energy Consumption', *Government Technology*, 27 July 2010. Online. Available at www.govtech.com/magazines/gt/How-to-Cut-Data-Center-Energy.html (accessed 10 April 2024).
8. See T. Bawden, 'Global Warming: Data Centres to Consume Three Times as Much Energy in Next Decade, Experts Warn', *Independent*, 23 January 2016. Online. Available at www.independent.co.uk/environment/global-warming-data-centres-to-consume-three-times-as-much-energy-in-next-decade-experts-warn-a6830086.html (accessed 10 April 2024). See also 'Fighting Climate Chaos', Greenpeace. Online. Available at www.greenpeace.org/usa/fighting-climate-chaos/click-clean (accessed 19 April 2024).
9. 'Cisco Global Cloud Index: Forecast and Methodology, 2016–2021 White Paper', updated 19 November 2018. Online. Available at www.virtualization.network/Resources/Whitepapers/0b75cf2e-0c53-4891-918e-b542a5d364c5_white-paper-c11-738085.pdf (accessed 10 April 2024).
10. See B. Panzica, 'Energy Consumption to Jump 53%', *Energy & Capital*, 20 September 2011. Online. Available at www.energyandcapital.com/articles/energy-consumption-to-jump-53-/72920 (accessed 10 April 2024).
11. See G. Reijn, 'Waarom kan het elektriciteitsnet vraag en aanbod niet meer aan?', *de Volkskrant*, 29 August 2019. Online. Available at www.volkskrant.nl/nieuws-achtergrond/waarom-kan-het-elektriciteitsnet-vraag-en-aanbod-niet-meer-aan~b692cfc2 (accessed 10 April 2024).
12. See E. Stuivenwold, 'The Netherlands Is European Leader in Data Centres', Dutch Data Center Association, 11 June 2019. Online. Available at www.dutchdatacenters.nl/en/nieuws/dutchdatacenters2019-2 (accessed 10 April 2024).
13. See 'Facebook Puts Zeewolde Data Centre Plan on Hold After Protests', *Dutch News*, 30 March 2022. Online. Available at www.dutchnews.nl/2022/03/facebook-puts-zeewolde-data-centre-plan-on-hold-after-protests (accessed 10 April 2024).
14. See Dossier- en ondernummer 32813 nr. 645, 'Kabinetsaanpak Klimaatbeleid', Tweede Kamer der Staten-Generaal, 16 December 2020. Online. Available at zoek.officielebekendmakingen.nl/kst-32813-645.html (accessed 10 April 2024).
15. See M. Meaker, 'These Angry Dutch Farmers Really Hate Microsoft', *Wired*, 30 March 2023. Online. Available at www.wired.co.uk/article/microsoft-netherlands-hyperscale-data-centers (accessed 10 April 2024).
16. See 'OM: Google-directeur kreeg smeergeld voor Eemshaven', *NOS Nieuws*, 24 November 2016. Online. Available at www.nos.nl/artikel/2144792-om-googledirecteur-kreeg-smeergeld-voor-eemshaven (accessed 10 April 2024).
17. See 'Kabinet besluit tot aanscherping regels hyperscale datacenters', Rijksoverheid, 16 February 2022. Online. Available at www.rijksoverheid.nl/actueel/nieuws/2022/02/16/kabinet-besluit-tot-aanscherping-regels-hyperscale-datacenters (accessed 2 August 2024).

Where They Hide the Clouds

Füsun Türetken

Not everything is metal, but metal is everywhere. Metal is the conductor of all matter... And thought is born more from metal than from stone. 1
– Gilles Deleuze and Félix Guattari, *A Thousand Plateaus: Capitalism and Schizophrenia*, 1983

Grounded in French philosophers Gilles Deleuze and Félix Guattari's notion of 'pan-metallism', this essay aims to foster novel perspectives in understanding the agency of matter – particularly metals. 2 Metals are agents that influence and register instances of (political) conflict. They are the conductors of all matter. Metals cross over into every organic form as salts and minerals, expressing a 'vital materialism' unprecedented in other substances. 3 Acknowledging metal's ubiquity and the aesthetic, cultural, socio-economic and political entanglements it creates, this essay examines metal's role in shaping the world of capitalist finance, belief systems, geopolitical relations and (digital) bodies.

The sky is constituted by invisible floating water and metallic particles. Weather clouds evoke a sense of weightless solidity. Sometimes they're opaque, sometimes translucent. Clouds are the sky in motion, fleeting from place to place. Clouds are a matter of life and death. One cannot escape the clouds. One cannot escape the sky. The word 'sky' conjures images of a lofty and unreachable dome, one distinctly separated from the Earth. This essay offers an alternative ontology – that the strata of the 'sky' as well as the digital cloud are not divided from the

Earth. Instead, they are constantly blended, both existing within and outside the other. Metals, extracted from Earth's crust and its inner depths, whether base, precious or rare, Earth's metals run the engines that form digital technologies and the cloud itself. The reterritorialization that occurs digitally influences representative democracies while reconstituting the relationship between democracy and society. The results of this are fragmented information societies, with social media channels, bubbles and echo chambers filled with fake news that undermine democratic procedures and democratic organs, leaving them in a parlous state.

Simultaneously, extractivist economies and mining hollow out landscapes leaving the planet's skeleton in a malady that I term 'Gaiaporosis': a conjoining of the disease *osteoporosis*, which leads to a deterioration of the bone structure in its density and minerals, and James Lovelock's *Gaia*. 4 The term 'porosity' is part of the vocabulary of biology and petrophysics, where it describes a void or space encapsulated in materials or rocks. This essay is based on the notion of a double bind of porosity: one in democratic procedures induced through technology and data, and another that occurs in earth, the biosphere and bodies, through the extraction of raw materials that are required to construct computers, the Internet and data centres aka 'the cloud'. We come to understand that it's only a small step from an etching by the Paper Architects discussed below, who resisted a project housing all information to today's Internet and cloud.
Its architecture is based on data centres and their smallest building block, the etched microchip processor or the circuit board. The cloud and its aforementioned ingredients colonize landscapes, bodies and minds. Insofar as data collection is the new model of capitalism, we thereby arrive at data capitalism, a new form of colonization. 5

Inducing Porosity: Etching Metal Surfaces

'Porosity' is a concept transferred from different disciplines such as biology, organic chemistry and the study of plants and animals. It indicates the existence of pores (from Greek *πορος*) which mean 'minute openings'. 6 Porosity defines the percentage of spheric hollow spaces (in a rock) by the total volume. It is written as either a decimal fraction between 0 and 1 – alike the 0s and 1s that define the binary code – or as a percentage. 7 Porosity is implemented here to understand what extractivist practices including digital technologies accelerate. The term 'dual porosity' refers to the conceptual idea that there are two overlapping openings but distinct bodies that interact. 8 Thus I argue here that a dual process of porosity is at work in two 'bodies' or systems that in fact overlap and interact.

Etching is a traditional printing technique: strong acid is applied to cut into the unprotected parts of a metal surface to make incisions in metal. Modern variants

60

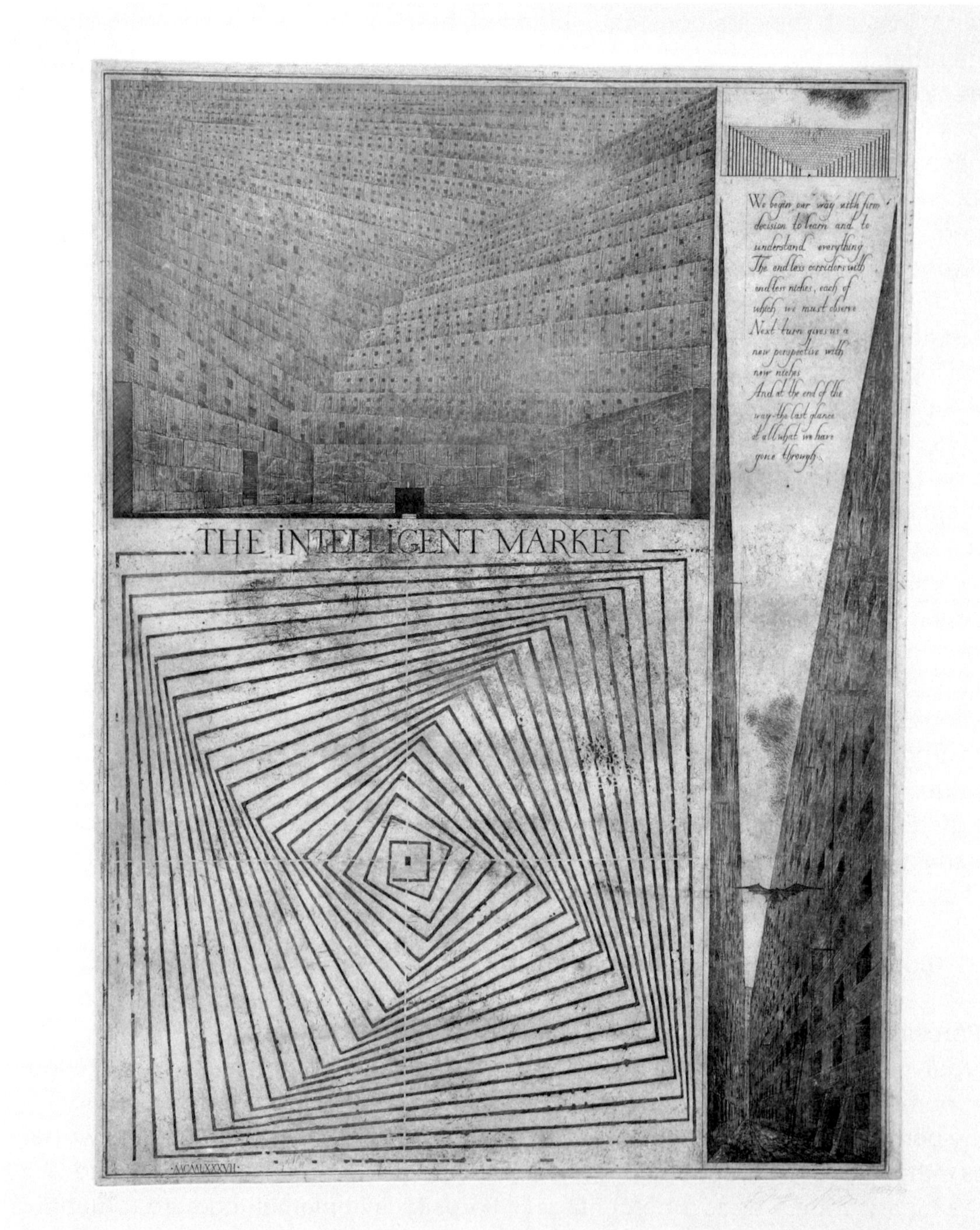

Alexander Brodsky and Ilya Utkin,
The Intelligent Market (Central Glass
International Architectural Design
Competition 1987).

such as microfabrication etching and photochemical milling are crucial techniques in producing circuit boards. In the image on the left, we find a connection between a print that is based on metal etching and its contemporary version, the circuit board or etched microchip processor. This technique of etching was applied by the artists Alexander Brodsky and Ilya Utkin, group members of the Soviet artists Paper Architects. 9 Their collaboration between 1978 and 1993 resulted in a criticism of the utilitarianism (in architecture) that was propagated by the Soviet Communist Party. Due to their reluctance to go along with the government's aesthetic principles, they were unable to build. Forced to materialize their ideas on paper instead, they deployed their designs to protest and entered competitions in the early days of Glasnost. 10 One of their etchings, titled *The Intelligent Market* from 1987, was an act of micro-resistance to a competition call that asked for a building 'devoted to the age of information, [to the] new opportunities supposedly opening up for humankind, where all this was taken to the extreme, [computer technology], where you were supposed to design a new kind of building [and] where you could get information in a second from anywhere in the world'. As the artists explain, they 'did not approve [of] ... all this overload of information – we didn't want to exalt it, which is what the organizers really wanted, [and] took the reverse position'. Their drawing '[a] tomb like architecture implies that knowledge needs to be exhumed from an obsolete era of obfuscation'. 11 It is reminiscent of today's architecture of data centres, modern mausoleums that enclose *data* related to human bodies while hermetically sealing their contents from the public eye. The Soviet call for 'a building offering all information at once from anywhere' sparks associative thoughts of a data centre, the unit that encapsulates both the Internet and the 'cloud'. The USSR's aspirations for a blueprint of such a building go hand in hand with its amplified interest in the data gathered by 'information ministries', as was recently depicted in Adam Curtis's film series *Russia 1985–1999: TraumaZone* (2022). The race between the United States and the Soviet Union peaked not only in the space race, but also in their mutual aspiration to build a national (civilian) computer network system. It is against this backdrop that *Intelligent Market* must be read, right at the threshold of the Iron Curtain, two years before the fall of the Berlin Wall and the end of the Cold War. 12

The Cloud That Never Was, Due to a Lack of Porousness

The astonishing history of the Soviet Internet that never came to be begins with the mathematicians, engineers and cyberneticists who prepared their own network. 13 Anatoly Kitov and Victor Glushkov 14 – inspired by Norbert Wiener and his publication *Cybernetics: The Human Use of Human Beings* (1950) – conceived of computerization as the solution to the Soviet Union's failing command economy and suggested establishing a state-automatized system to gather and operate data, namely OGAS (Общегосударственная автоматизированная система учёта и обработки информации [National Automated System for Computation and Information Processing]). 15 The implementation of computers as a planning instrument was politically controversial in the 1960s. In *How Not to Network a Nation: The Uneasy History of the Soviet Internet* (2017), Benjamin Peters examines one of the fundamental problems that troubled the Soviet experiment: information. 16 Needless to say, the underlying method behind state socialism is the centralized planning of a government, its decision making acquires reliable data. The greatest challenge for such a government is then the assessment of all this gathered data, especially without computer technology. 17 'The biggest advantage that the United States wielded over the USSR appears to have less to do with the market independence of private commerce than the *porousness* of research, resources and knowledge flows between military and civilian projects.' 18 As Orit Halperin writes: 'At this pivotal moment, demarcated by a catastrophic world war, these sciences were part of producing an aspiration for a new world constituted of information; but not without producing a novel set of conflicts, desires and problems.' 19 Finally in 1970, the Politburo ended the program OGAS without realizing it – one year after the American ARPANET went online. 20 Today, 53 years later, the concept of the InterNjets (No to the Internet) surfaces in today's Russia, and the open Internet is in turmoil. 21 The Internet Research Agency at 55 Savushkina Street in St Petersburg is one example of these networks operating in the dark (out of sight), it is an enterprise that influenced the 2011 elections in Russia and was 'indicted by special counsel Robert Mueller on charges of conspiring to tamper in the 2016 US presidential campaign' and is 'the epicenter of a propaganda machine', where the 'internet and state media merged into one'. 22 In the light of fragmentation of digital communication networks and of society, Cass R. Sunstein has claimed that 'diverse speech communities' are created 'whose members talk and listen mostly to one another. ... Groups will tend to polarize in a way that can breed extremism and even hatred and violence. New technologies, including the Internet, are creating these echo chambers.' 23 Sunstein is concerned with how rumours may impair the effectiveness of government and undermine its legitimacy. Hannah Arendt refers to something similar in *The Human Condition* (1958), when she writes that 'speech has lost its power'. She addresses the loss of the capacity to act politically, and the problematic representation that comes with automation powered by machines, swarms and clouds. 24 'The networking between nervous nets and neurotic governments supports the argument by Patricia Clough that we live in an affective economy, where the social and the economic emerge from the manipulation and circulation of sentiments and sensation before consciousness.' 25

Conclusively, Vladimir Putin's infrastructural plans for a closed Russian net titled Runet in part returns to the idea of competing nation-states. 26 What is desired is not the networking of nets, as realized by ARPANET for the first time in 1969, but their direct political control. As Sebastian Giessmann puts it: 'The reterritorialization of digital nets is the geopolitical pattern of the 2020s.' 27 Once again nation-states set the benchmark, as they did 50 years ago, in the contestation over digital sovereignty and artificial intelligence. 28 One could conclude that this is contrary to what constitutes porousness.

Hiding the Cloud

Brodsky and Utkin's refusal in 1987 to create a building to house all information can be interpreted either as prophetic, or as a form of naive, yet critical resistance that could not withstand the events of time. In the meantime, the etched circuit boards and the cloud itself, together with its critical infrastructure, have landed on planet Earth. In its states of aggregation from liquid, gaseous to solid form, the cloud has only further solidified territorial claims. Architectural theorist Pier Vittorio Aureli states, 'Any power, no matter how supreme, totalitarian, ubiquitous, high-tech, democratic, and evasive, at the end has to land on the actual ground of the city and leave traces that are difficult to efface.' 29 Still, attempts to erase or camouflage those traces exist. In 2008, before the launch of its 'iCloud', Apple attempted to hide the location of a data centre from satellite images provided by Google. Instead, images of a blank space or a Greenfarm were shown. 'We don't know how Apple kept Google from revealing its latest project, but after iCloud was announced [by Steve Jobs], the curtains were peeled back' and the image of the base of Apple's iCloud came to digital life. 30 'We're going to move the digital hub, the center of your digital life, into the cloud,' Jobs said. 'Some people think the cloud is just a big disk in the sky. We think it's way more than that.' 31 What appeared was a ghost-like nondescript building attending only to itself, which is one of the distinctive features of these data centres, since not many employees are required for their maintenance. What they instead require are cheap electricity and power, the reasons why this building is in Maiden, a town in North Carolina, where coal is plentiful. 32 The refusal by technological companies to transparently publish information on their data centres is often justified by recourse to security, as the building itself as well as their information policies remain opaque and hermetic. The rise in the artificialization of soils and natural spaces across the northern hemisphere – as depicted by Niels Schrader and Roel Backaert in this very book, *Acid Clouds* – as well as in such a tiny country as the Netherlands has resulted in land being sealed in order to build data centres and other infrastructures necessary for the acceleration of the digital economy. In *Server Manifesto: Data Center Architecture and the Future of Democracy* (2022) Niklas Maak 'points to [this] physicality of an infrastructure that is otherwise presented to us as being as impalpable and evanescent as a cloud'. 33 Given the socio-economic and political importance of data ownership, it is also of strategic

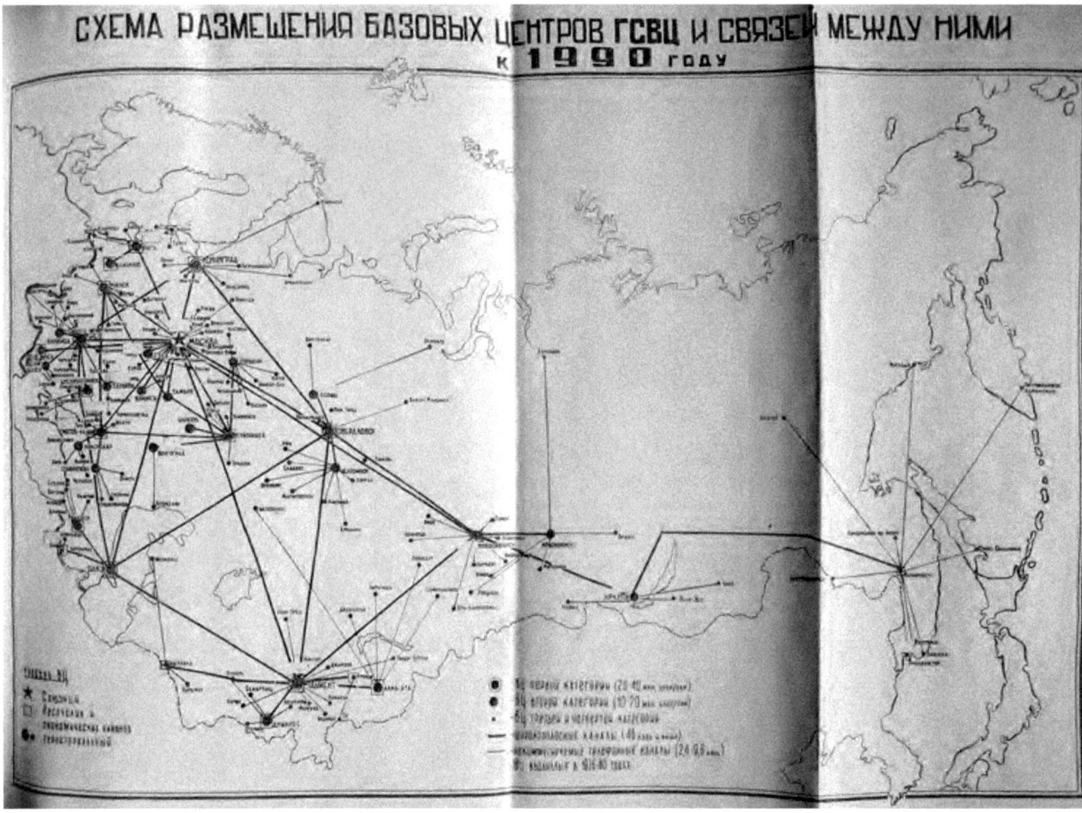

Map of the State Network of Data Centers
EGSVT, Plan for the Year 1990, 1964
in B. Peters, *How Not to Network a Nation: The Uneasy History of the Soviet Internet*
(Cambridge, MA: The MIT Press, 2017), 112.

interest that data centres often tend to be anonymous, placed either on the outskirts of cities or now colder regions. Their logic seems to be: out of sight, out of mind. Apple's attempted vanishing act resembles the story of Stalin building the city Sillamäe in the late 1940s to then erase it from the map: '[T]he scientists worked on top-secret metallurgical processes, refining uranium, rare earths, niobium and tantalum.' 34 Both examples recall architect and activist Eyal Weizman's 2015 essay 'Violence at the Threshold of Detectability', in which he describes the asymmetry in information in terms of the resolution, of pixels and metres, that a government holds in image technologies, e.g. satellite imagery, relative to the individual citizen. 35 Can we assume that data centres will simply slip through the pixels on any contemporary map? As the design duo Metahaven write:

> A data center is a windowless, large, flat building... [T]he precise locations of data centers are secret. To users, the cloud likewise seems ... 'formless cyberspace' indeed... [Its] physical form [is] concrete infrastructure. If the enormous, energy-slurping data factories are the cloud's true form, then ... the[y] recall the medieval castle, ... the military base, ... [and] the political and military conflicts that have dominated geography since recorded history. 36

Dissimulating Metals

There are 17 rare earths in total, including those metals grouped under the name *lanthanides*. The word 'lanthanum' comes from the Greek meaning *lanthanein* 'to hide, to lie hidden'. 37 And this is exactly what rare earth metals do, since they never occur in their pure form, but only in combination with other elements. When one investigates what 'data' is made of, we encounter the same obstacle: hidden information. Not only do our metallic actors hide, but the Big Tech companies also hide. This hide and seek game by companies and political lobbies with societies proves perilous. We can then ask: what are data centres made of? They consist of the same basic computing as our personal computers. Meanwhile, the hardware components that make up the IT infrastructure of a data centre include the same precious metals applied in servers and storage equipment, plus additional infrastructure to support them such as power and cooling systems. What hardware precisely is in data? How much precious metal is in a circuit board? A server, the motherboard, and the circuit board usually contain 30 per cent plastic, 30 per cent inert oxide and 40 per cent metal. 38 To break it down, the building block essential for the data centre is the microchip. A company called ASML, based in Eindhoven, is the market leader in factoring machines that produce microchips. Due to this fact, the firm is of geopolitical relevance, since chips have become a strategic asset. ASML, formerly an obscure company, has fallen prey to meddling by the major powers. It isn't the first time in history of course that metal-etched surfaces have been key to our economies, as we can think of coins. When looked at carefully, a microchip itself is a complex architecture that resembles our etched collage of the Paper Architects. As Jos Benschop, the senior vice president technology at ASML, explains: 'If you put a chip under a microscope you see tiny chambers with zeros and ones. A pattern of tracks, a 3D architecture, structure with rooms and hallways, where the information is stored at the bottom.' 39 Its labyrinthian geometry resembles the design by Brodsky and Utkin. In hindsight this miniature architecture encapsulates an amount of data so extensive, that it surpasses the aspirations of any super-power 40 years ago. Millions of microchips and circuit boards with their internal etchings store information, and this is what clouds are made of. Small molecular chips. Once we look closer at the very bottom of these chips, we understand what data centres are really *made of* and *for*: that is, information extracted from humans. Displacing human experiences into little zeroes and ones and pressing them into silicon, plastic and metal. Thus, we need to look beyond the surface of materials, and beyond the notion of the extraction of raw material and conflict minerals. Like the *lanthanides*, (data) extractivism lies hidden, it never occurs in its pure form, but solely in combination with other elements.

If we wish to understand the connection between computation and material extractivism it is helpful to turn to a little arithmetic calculus of ingredients. Today, most computers and electronic devices contain circuit boards, which consist of various types of precious metals including gold, palladium, platinum and silver. 40 High levels of precious metal content are also to be found in printed circuit boards. What amounts here is a pyramid of metals. An overview below allocates valuable and precious metals in computers. Moreover, it demonstrates why contemporary societies have been for centuries, and are still, on a quest for metallic resources. 41 The current (strategic) battle between nation-states over rare earth metal resources is related to the dependence on (these) metals, which illustrates the costs attached to our technological and economical affordances. If we think of a metallic genesis we arrive at the first exploding supernova, but if we think of a genesis of the ingredients involved in contemporary computer systems we find that those used in the period of the Space Race in the 1960s contained more gold due to the need for corrosion resistance and requirements of higher quality, 'with the majority of components manufactured specifically to support the manned space missions and national defense projects'. 42 Those electronic circuit boards and components often used generous amounts of gold plating. After 1970, the electronic industry reduced the amount of gold plating due to the rapidly rising cost of gold. Due to these circumstances, alternative precious metals were explored. With the acceleration of cloud computing we arrive at a recipe of ingredients that registers the slow transition to the Rare Earth Metal Age.

Aluminium	Printed circuit boards, computer chips, hard drives, CPU heat sinks
Copper	CPU heat sinks, wiring and cables, printed circuit boards, computer chips
Cobalt	Hard drives
Gold	Printed circuit boards, computer chips (CPU), connectors / fingers
Neodymium	Hard drives (magnets)
Nickel	Circuit board components
Palladium	Hard drives, circuit board components (capacitors)
Platinum	Hard drives, circuit board components
Silver	Printed circuit boards, computer chips, keyboard membranes, some capacitors
Tantalum	Circuit board components (some capacitors)
Tin	Printed circuit boards, computer chips
Zinc	Printed circuit boards

Another way to read the ingredients is the highest precious metal content by weight in computer components:

Gold / Silver / Palladium	Computer CPU (processor)
Gold	Memory (RAM) and Circuit board fingers, connectors, pins
Gold / Silver	Circuit boards (Motherboards)
Palladium / Platinum (on HD platters)	Hard drives
Palladium / Silver (Capacitors)	Discrete electronic components
Gold / Silver / Platinum / Palladium	Whole computers

Aluminium, Copper, Cobalt, Gold, Neodymium, Nickel, Palladium, Platinum, Silver, Tantalum, Tin and Zinc seem to then be the main actors in the age of computation. From the printed circuit board to the connector, to the server to the data centre. These minerals work together in machinic assemblages unfolding their specific qualities and their conductive qualities to store or to project our streamed films and photographs from the cloud. Sometimes in darker moments they accelerate the navigation of a laser, to target and fire those weapons attached to unmanned vehicles and drones that swarm the skies. Returning back to the skies and clouds we know that metallic molecules colonize clouds. The sky has not only become metallic but foremost vibrant. 43

Gaiaporosis

The controversial neo-reactionary (and one might dare to state white supremacist) philosopher Nick Land developed the concept of 'geotrauma' in the late 1990s to radicalize psychoanalysis beyond what Deleuze and Guattari had reimagined in *Anti-Oedipus*, with their concept of schizoanalysis. 44 Land materialized the unconscious all the way into the body of the Earth. 45 This concept of geotrauma can be understood as the geological formation of the Earth as a psychoanalytical process. 46 The layers of sediments that are peeled back in the process of mining for rare earth metals (REM) and precious metals leave a perforated landscape behind. Layer after layer, vectors of violence intersect with vectors of power; hollowing out the skeleton of Earth's body without 'care', capital leaves behind a planet suffering from 'Gaiaporosis'. 47 With this notion I refer to the extractivist tendencies and the hollowing out of planet Earth's raw materials, landscapes and biosphere. Nowadays the colonizer comes disguised as a corporation or development agency. The ramifications of conflict minerals, mining's hazardous and toxic working conditions, and the resulting impact on the bodies of the workers remind us that the precarious neocolonial working conditions match those of the past. The system of exploitation (and slavery) causes physical and psychological pain, and the labourer's trauma of slavery, deeply inscribed into history and their bodily memory, is revived once again. 48

Living Minerals

Sociologist Nick Couldry argues that capitalism today is driven by the notion of gathering data as an economic imperative. 49 As a result, 'human experience ... is becoming the target of profitable extraction'. This indicates that colonialism is reemerging as data colonialism. 50 Digital policy warrior Anri van der Spuy asks: 'If the territories of the new data colonialism are you and me and our very bodies as generators of data, who and where are today's Dutch East India Companies (as perpetrators of data colonialism)?' 51 The dimensions of exploitation can only be understood through the relation of humans to their data, data centres and tech giants. Sociologist Vincent Mosco claims that the ethereal dimension of the 'cloud' needed to operate external data storage and data management stands in contrast to its physical material realities and the materials extracted from the Earth's crust, which are needed for the operation and the expropriation of entire populations. 52 Writers Sandro Mezzadra and Brett Neilson have emphasized how extractivist practices such as the literal hollowing out of planet Earth and the excavation of its biosphere to gain resources are connected to (forced indigenous and slave) labour. 53 If we consider technologies as geological processes, we comprehend the exploit of non-renewable deep-time resources (that have formed over millions of years) that are needed to operate contemporary technologies, many of which are short-lived smart devices with built-in obsolescence. 54 One needs to fathom extractivism in a larger scope, from mining to dispossession, to data extraction, to labour and body extraction, biocapitalism extraction. 55 In the Capitalocene and the Anthropocene, 'extractivism has become a core symptom of the planetary disease of late capitalism ...' 56 As Achille Mbembe argues, 'the connection between forced labor and extraction ... allows an understanding of the slave trade in extractivist terms'. 57 He writes that African people are 'transformed into living minerals from which metal is extracted, giving rise to a transition from 'homme-minerai to homme metal and from homme-métal to homme-monnaie'. 58 Based on this notion by Mbembe, and Deleuze and Guattari's argumentations in *A Thousand Plateaus*, I argue that forced labour is always connected to a human and a body that has become metallic – and can be coupled into endless metallic formations. 59

Metallic Bodies and Superconductive Biopolitics

In *A Thousand Plateaus* Deleuze and Guattari state: 'There is no biosphere or noosphere, but everywhere the same Mechanosphere.' 60 The term refers to a planet-wide intermeshing of mechanical, architectural and biological processes – the most notable being how the technosphere started to invade the biosphere, with catastrophic climate conditions as a result. The concept informs much of Deleuze and Guattari's philosophical thinking, since it is this idea that advances their 'machinic materialism' and explains their notion of panmetallism. It is their attempt to materialize psychoanalysis, and to read it through the canon of pragmatist materialism; by conceptualizing the body as an assemblage of machines coupled together, the Freudian id becomes the machinic it. The metallic body is birthed from this logic like Athena from the head of Zeus. The cybernetic instantiation, the current computational age relies heavily on REMs to act as superconductors in (semi-)intelligent systems. 61 Their superconductive properties are the sustaining lifeblood of machinic metallic bodies. Following the logic of fiction at large we arrive at an alien totality of walking clouds, and at the notion I coin 'digital colonialism'[62] or 'techno-colonialism': the 3D digital supermodel named Shudu who presents an ideal example of REM-based exploitation of the black female body and of new formations that come into being when metal meets humankind.[63] The reliance on metals, for example, contextualizes contemporary and future conditions of metal's complicity in labour.[64] From the Fordist proletariat's brute force to the post-Fordist condition of immaterial labour exploitation and the current demand for REMs, we can highlight metal's complicity in power systems, which I call its 'political conductivity'. REMs are often superconductive.[65] The superconductivity allows for a more intense and faster electrical computation, which is necessary for artificially intelligent computer chips. Superconductive materials are mostly made of REM elements; thus, artificial intelligence (AI) and other silicon-based technologies are materially reliant on the terrestrial abundance of these elements. It is this shift in metal usage that has so affected labour conditions and bodies. The agency of rare earth metals in their reign over bodies relies on the infrastructures and technologies. These elements have gained prominence as a critical strategic resource, in their role in the civil sector, in high-tech industry and in defense technology. Their geopolitical significance is apparent since they are assumed to be the 'next oil', but also in the dispute over resources between nation-states.[66] David S. Abraham dedicated his book *The Elements of Power* to rare earths that 'make products smaller, lighter, faster, more powerful, thinner, brighter, stronger'.[67] He explains that more 'rare earth magnets used in hard drives in cloud data storage centers [are needed]'.[68] They are the nodes of the network essential to operate digital lives. Neodymium, one of the most powerful permanent rare earth magnets, used in computer hard disks and other devices: is the other building block for all of the data centres that colonize our contemporary landscapes.[69] Following Reza Negarestani's logic, I argue that rare earth metals use capitalism to leave their tellurian subterranean existence. Once unearthed, they will come to life as commodities in different states, either 'malleable (indium), ductile (niobium), toxic (cadmium), radioactive (thorium), or magnetic (cobalt [and neodymium])' as Abraham states, and 'like characters in the X-Men comics, they all have their own superpowers'.[70] With data centres moving to the Global North and the Arctic territories, e.g., Iceland, Greenland and Sweden, we will soon see that these architectures will incur on the very territory that provides all ingredients necessary in the first place.[71] A leap back into Cold War times shows us that both the military programs of the Unites States as well as the Soviet Union investigated how to access the sea around the North Pole, even considering to melt the caps in order to access resources and trade routes.[72]

Starting from buildings that house information, etched into metal surfaces, we have traversed a cloud that never came into being – the camouflaged iCloud – and arrived at information hidden from our eyes. If we look, we can see how clouds are coming to darken our skies and casting shadows over our soil and land. The metallic cloud has ushered in an age of obfuscation. The rain of information that clouds supposedly provide can be of a misleading nature, as more data doesn't equal more transparency. Information based on metals works its way into all strata. From the mining of stars to metals that are dug up from Earth's substrata. The world of the cloud depends on this vertical stretch, on the vast network and geometry of data centres located around the globe, with their building blocks such as the microchip processors and super magnets. The panmetallic world is structured by metals in all states. They are the protagonists of events and architectures, and they keep digital figurines alive. They take shape as active agents. The cloud drags behind it an architectural footprint that provides the energy needed while leaving an imprint of Gaiaporosis by reaching into the bowels of planet Earth. A metal fever has befallen us. A metallic virus hollows out Earth, Gaia, democracies and human bodies only to keep digital bodies and their houses, the data centre, alive. Metals work in these machines. All of which conduct and form the cloud. Metals are everywhere; even in the space between metals they are the vectors of power.

1. G. Deleuze and F. Guattari, *A Thousand Plateaus: Capitalism and Schizophrenia* (Minneapolis: University of Minnesota Press, 1983), 411.
2. Ibid.
3. See E. Jones, 'Meditation on Metallurgy', Ganoksin Jewellery Making Resources. Online. Available at *www.ganoksin.com*/article/meditation-metallurgy (accessed 28 June 2023).
4. J. Lovelock developed the Gaia theory in the 1960s. He suggested that our planet self-regulated like a living organism. Later he was joined by Lynn Margulis in this conceptual thought. Here, I graft a word from gaia, which means Earth and life in ancient Greek; and osteoporosis, the illness of the bone, resulting in the word combination gaia-porosis. Porosity derives from different disciplines and describes an irregular void or space in a solid material (rocks).
5. A. van der Spuy, 'Colonising Ourselves? An Introduction to Data Colonialism', *Media@LSE blog*, 19 March 2020. Online. Available at *blogs.lse.ac.uk*/medialse/2020/03/19/colonising-ourselves-an-introduction-to-data-colonialism (accessed 28 June 2023).
6. S.D. Kotsopoulos, 'Design Concepts in Architecture: The Porosity Paradigm', in: D. Allemang et al. (ed.), *Semantic Web and Web 2.0 in Architectural, Product and Engineering Design* (2007). Online. Available at *www.ceur-ws.org*/Vol-294/paper08.pdf (accessed 28 June 2023).
7. 'Understanding Porosity and Density', Wisconsin Geological and Natural History Survey. Online. Available at *wgnhs.wisc.edu*/maps-data/data/rock-properties/understanding-porosity-density (accessed 28 June 2023). The term porosity is used in multiple fields including biology, ceramics, Earth sciences, medicine, metallurgy, organic chemistry, petrophysics, pharmaceutics and soil mechanics among others.
8. 'Porosity', Wikipedia. Online. Available at *en.wikipedia.org*/wiki/Porosity (accessed 28 June 2023).
9. K. Hardiman, 'Four Brodsky and Utkin Etchings and Rauschenberg's Fresco Contest', *Rauschenberg Collecting & Connecting*, Shuffle: A Rauschenberg Artwork Lending Library, 2014. Online. Available at *shuffle.rauschenbergfoundation.org*/exhibitions/nasher/essays/Hardiman_four-brodsky-and-utkin-etchings (accessed 28 June 2023). The etchings by Brodsky and Utkin were influenced by Egyptian tombs, Ledoux's and modern architecture, and other historical precedents. K.A. Conover, 'Architecture: From Flat Fantasies to Real-World Designs: Free of Communist Constraints, "Paper Architects" Move to Three Dimensions', *Washington Post*, 12 May 1994, E13. The second correlation here is the medium and word *paper*: when we think of Computer Centers in the 1970s we encounter a world that is made of paper. From the punch card, a storage medium, to printed paper. Historically, the punch card was part of the start of electronic data processing. The first punch cards by Herman Hollerith were used in a census in the United States held over one hundred years ago in 1890. E. Zopfi, 'A World Made of Paper: Computing Centers in the 1970s', in: M. Dommann, H. Rickli and M. Stadler (ed.), *Data Centers: Edges of a Wired Nation* (Zürich: Lars Müller, 2020).
10. Hardiman, 'Four Brodsky and Utkin Etchings and Rauschenberg's Fresco Contest', op. cit. (note 9).
11. Ibid.
12. Glasnost represented a policy of openness in the activities of state institutions and freedom of information. In Russian it has referred to 'openness and transparency' since the eighteenth century.
13. S. Giessmann, 'Wie die Sowjetunion beinahe das Internet erfunden hätte', *Frankfurter Allgemeine Zeitung*, 4 October 2010. Online. Available at *www.faz.net*/aktuell/wissen/computer-mathematik/geschichte-des-internjets-wie-die-sowjetunion-fast-einmal-das-internet-erfunden-haette-16976408/in-novosibirsk-hatte-die-16981596.html (accessed 4 November 2020).
14. Viktor Mikhailovich Glushkov was vice president of the Ukrainian Academy of Sciences and Director of the Institute for Cybernetics in Kiev until 1982.
15. OGAS is the Soviet counterpart of ARPANET. It is an example of how network projects could have developed in societies that were not preoccupied with markets, democracies and personal liberties. Using Hannah Arendt's vocabulary of 'oikos' and 'polis' to discuss how the heterarchical but closed Soviet command economy felt threatened by the idea of a rigid algorithmic system, Benjamin Peters redefines the word 'technology' as the study of the crafts of social relations. See B. Peters, *How Not to Network a Nation: The Uneasy History of the Soviet Internet* (Cambridge, MA: The MIT Press, 2017).
16. See B. Peters, *How Not to Network a Nation: The Uneasy History of the Soviet Internet* (The MIT Press, 2017).
17. Giessmann, 'Wie die Sowjetunion beinahe das Internet erfunden hätte', op. cit. (note 13).
18. Ibid. In the USSR although academic institutions were similarly connected to each other, as in the United States, the Soviet military considered any further expansion of civilian activity wasteful, hindering the growth of their Internet system. I am interested in how porousness, which I will go on to note as a hollowing out, undermines democratic procedures, systems or materials, and (planetary) bodies.
19. O. Halperin, *Beautiful Data: A History of Vision and Reasons Since 1945* (Durham, NC: Duke University Press, 2015), 41. Halperin states that the word 'cybernetics' suggests a history in governance, and that Wiener explained that the word cybernetics and government have the same route in Greek.
20. The two network projects were different: OGAS aspired to network the entire industrial base and increase the efficiency of the command economy with a computer network. It included all macro-economical activities, models and statistics. The ARPANET, the predecessor to the present-day Internet, in contrast, created connections between scientific institutions.
21. Giessmann, 'Wie die Sowjetunion beinahe das Internet erfunden hätte', op. cit. (note 13).

22. A building housing a fake news division, bloggers and social media commentators as well as a graphics department creating memes. In creating a 'circle of lies' – a feedback loop where troll postings reinforced news on state media in a former Russian election slowly 'the internet and state media merged into one'. C. Maynes, 'Inside the Internet Research Agency: A Mole Among Trolls', *VOA*, 17 April 2018. Online. Available at www.voanews.com/a/inside-the-internet-research-agency-a-mole-among-trolls/4352107.html (accessed 28 June 2023).
23. Cass R. Sunstein, *Republic.com 2.0* (Princeton and Oxford: Princeton University Press, 2007), 44.
24. Ibid.
25. Halperin, *Beautiful Data*, op.cit. (note 19), 184; T. Terranova, 'Another Life: The Nature of Political Economy in Foucault's Genealogy of Biopolitics', *Theory, Culture and Society* (2009) vol.26, no.6, 235.
26. Runet is an online Russian-language community. The term was coined by blogger Raffi Aslanbekov in 1997.
27. Giessmann, 'Wie die Sowjetunion beinahe das Internet erfunden hätte', op.cit. (note 13).
28. Ibid.
29. Metahaven, 'Captives of the Cloud: Part II', *e-flux journal* (October 2012), no.38. Online. Available at www.xenopraxis.net/readings/metahaven_cloud2.pdf (accessed 9 March 2023).
30. N. Jackson, 'Apple's Splashy New Data Center Finally Appears on Google Earth', *The Atlantic*, 1 June 2011. Online. Available at www.theatlantic.com/technology/archive/2011/06/apples-splashy-new-data-center-finally-appears-on-google-earth/239750 (accessed 9 January 2023).
31. R. Miller, 'Steve Jobs Provides a Look Inside the iDataCenter', *Data Center Knowledge*, 6 June 2011. Online. Available at www.datacenterknowledge.com/archives/2011/06/06/steve-jobs-provides-a-look-inside-the-idatacenter (accessed 26 February 2023).
32. See T. Prickett Morgan, 'A Peek Inside Apple's iCloud Data Center', *The Register*, 9 June 2011. Online. Available at www.theregister.com/2011/06/09/apple_maiden_data_center (accessed 26 February 2023). Apple has increased its emphasis on consumer cloud services and in turn data centres over the past years. The company builds the largest end-user solar farm and engages in non-fuel energy gain for its data centre in Maiden, North Carolina. P. Jones, 'Apple Confirms Solar Farm at Maiden Data Center', *Datacenterdynamics*, 21 February 2012 (accessed 26 February 2023). Online. Available at www.datacenterdynamics.com/en/news/apple-confirms-solar-farm-at-maiden-data-center (accessed 28 June 2023).
33. Regine, 'Server Manifesto. Data Center Architecture and the Future of Democracy', *We Make Money Not Art*, 15 August 2022. Online. Available at www.we-make-money-not-art.com/server-manifesto-data-center-architecture-and-the-future-of-democracy (accessed 26 February 2023). A crucial aspect is that the geolocation of data centres doesn't matter, in order to form the cloud and the network they can be dispersed around the globe.
34. Sillamäe town officials interviewed by D.S. Abraham, Sillamäe, Estonia, 22 January 2013 in D.S. Abraham, *The Elements of Power: Gadgets, Guns, and the Struggle for a Sustainable Future in the Rare Metal Age* (New Haven, CT and London: Yale University Press, 2015), 67–68.
35. E. Weizman, 'Violence at the Threshold of Detectability', *e-flux journal* (April 2015), no.64. Online. Available at www.e-flux.com/journal/64/60861/violence-at-the-threshold-of-detectability (accessed 1 December 2022).
36. Metahaven, 'Captives of the Cloud II', op.cit. (note 29); see also P.V. Aureli in P.V. Aureli, B. Groys, Metahaven and M. Vishmidt, 'Form', in: Metahaven and M. Vishmidt (ed.), *Uncorporate Identity* (Baden: Lars Müller, 2010), 262.
37. 'Lanthanide', *OED2* 1.14 (1994) on CD-Rom, OUP.
38. H. Zang, 'How Much Money Can You Make by Recycling Waste Circuit Boards?', LinkedIn, 7 November 2019, Online. Available at www.linkedin.com/pulse/how-much-money-can-you-make-recycling-waste-circuit-boards-zhang (accessed 1 March 2023). The main direction of recycling the waste circuit board is metal recovery. The metal contains approximately 20 per cent copper, 8 per cent iron, 2 per cent nickel, 4 per cent tin, 2 per cent zinc, and precious metals such as 0.1 per cent gold, 0.2 per cent silver and 0.005 per cent palladium.
39. Interview with J. Benschop, 'ASML's Secret', *VPRO Backlight*, 12 September 2022. Online. Available at www.vpro.nl/programmas/tegenlicht/kijk/backlight/asmls-secret.html (accessed 28 June 2023).
40. 'Precious Metal in Computer Components', *Concept Management*. Online. Available at www.conceptmanagement.co.uk/knowledge/materials/precious-metal (accessed 28 June 2023).
41. Compared to old systems that were designed for any application in the military realm such as aerospace defense, radar systems hold the highest amounts of gold compared to contemporary commercial electronics. In a Department of Defense study for recycling surplus government equipment and for reducing precious metals use in government systems, computers from the 1970s were found to have contained significantly more gold in them as ones manufactured later. This reduction in the early 1980s derived from the development of smaller electronic components needing less gold. Online. Available at www.chipsetc.com/valuable-metals-used-in-computers.html.
42. Valuable Metals used in Computers. Online. Available at www.chipsetc.com/valuable-metals-used-in-computers.html (accessed 28 June 2023); with typical thicknesses in the range of 3 to 8 μm, with some applications being as thick as 20 μm. A micron (μm) is 0.001 of a millimeter.

43. In my text 'Colonizing Clouds' soon to be published, I refer to a metallic sky that is vibrant. The planet's ephemeral most delicate zone, the sky, is neither empty nor made of pure air, oxygen, nitrogen, argon, helium or any other serene matter. It is a vessel without boundary, filled with elements originating in earthly and other celestial bodies. The sky vibrates with a living composition – from ancient star-matter to microbes that mysteriously thrive in the higher atmosphere. This vibrant sky is filled with excitable matter that is constantly mingling, coupling, bouncing in a frenetic dance to create the phenomena that we experience as weather. Clouds are discussed not as 'things' or objects but rather phenomena, drawing on Karen Barad's argument that phenomena are the ontological inseparability of agentially intra-acting 'components'. Here, minerals such as antimony, arsenic, cadmium, copper, iron, lead, selenium, thalium, tellurium, tungsten, magnesium and zinc, shot into the stratosphere by volcanic eruptions, meet the unearthed materials released by humans and colonize the different strata of the sky in vertical succession.
44. See N. Land, *Fanged Noumena: Collected Writings 1987–2007*, R. Brassier and R. McKay (ed.) (Falmouth: Urbanomic, 2018). Materialists Deleuze and Guattari insisted that Freud's way of looking at mental health left the 'socius' out of the equation. Thus, schizoanalysis was the attempt to view mental health in terms of sociopolitical struggle. In other words, while Freud claimed that the neuroses stem from the desire to murder a parent and to have sexual relations with the other, schizoanalysis claims that, if mental illness is not a product of capitalist desire-oppression, it is at least amplified by it.
45. It is worth clarifying how Land's work has been used. It does not mean that I agree with his political positions. Put simply, the Land of the 1990s is responsible for a quite different *oeuvre* than is seen in his contemporary work.
46. Moreover, he claims that the social surroundings of a specific agent (be it human or nonhuman) are what affects subjective mental health. These scarred landscapes cause humans ecological grief, that is, a grief that comes with traumatic events such as volcanic catastrophes or climate change.
47. M.P. de la Bellacasa, *Matters of Care: Speculative Ethics in More Than Human Worlds* (Minneapolis: University of Minnesota Press, 2017). The idea and term of 'Gaiaporosis' was developed in collaboration with Natalie Konopelski.
48. Toni Morrison explains that slavery took a toll on the white (man's) psyche: 'Slavery broke the world in half, it broke it in every way. It broke Europe. It made them into something else, it made them slave masters, it made them crazy. You can't do that for hundreds of years and it not take a toll. They had to dehumanize, not just the slaves but themselves. They have had to reconstruct everything in order to make that system appear true.' T. Morrison in P. Gilroy, 'Living Memory: An Interview with Toni Morrison', *Small Acts: Thoughts on the Politics of Black Cultures* (London: Serpent's Tale, 1993), 178.
49. A. van der Spuy, 'Colonising Ourselves? An Introduction to Data Colonialism', London School of Economics, 19 March 2020. Online. Available at *www.blogs.lse.ac.uk/medialse/2020/03/19/colonising-ourselves-an-introduction-to-data-colonialism/#_edn2* (accessed 28 June 2023).
50. Ibid.
51. N. Couldry and U. Mejias, *The Costs of Connection* (Stanford, CA: Stanford University Press, 2019), xi.
52. See V. Mosco, *To the Cloud: Big Data in a Turbulent World* (Boulder, CO: Routledge, 2014).
53. S. Mezzadra and B. Neilson, 'On the Multiple Frontiers of Extraction: Excavating Contemporary Capitalism', *Cultural Studies* (2017) vol. 31, no. 2–3, 185–204. Online. Available at *www.datafarms.org/wp-content/uploads/2023/02/Mezzadra-Neilson-Extraction-2017-Accepted-Version.pdf* (accessed 28 June 2023)
54. Ibid.
55. Extractivism as a critical concept has an expansive meaning, encompassing wide aspects of capitalism and forms of exploitation. It includes the economic, political, cultural and ecological problems, largely displacing capitalism itself, while it encapsulates production, exploitation, environmental destruction, digitalization and new 'ontological assemblages'. See J.-A. McNeish and J. Shapiro, 'Introduction', in: J.-A. McNeish and J. Shapiro (ed.), *Our Extractive Age: Expressions of Violence and Resistance* (London: Routledge, 2021), 3; A. Dunlap and J. Jakobsen, *The Violent Technologies of Extraction: Political Ecology, Critical Agrarian Studies and the Capitalist Worldeater* (Cham, Switzerland: Palgrave Macmillan, 2020), 34, 100, 120–121; J. Bellamy Foster, 'Extractivism in the Anthropocene: Late Imperialism and the Expropriation of the Earth', *Science for the People* (Autumn 2022) vol. 25, no. 2. Online. Available at *www.magazine.scienceforthepeople.org/vol25-2-bleeding-earth/extractivism-in-the-anthropocene/#easy-footnote-bottom-16-15637* (accessed 28 June 2023). Biocapitalism refers to processes of the primary valorization of materials derived from human bodies and nonhuman living beings, to the meaning of these processes for capitalist accumulation strategies and to related transformations of modes of labour, exploitation and subjectivation. S. Lettow, 'Biocapitalism', *Krisis: Journal for Contemporary Philosophy* (2018), no. 2. Online. Available at *www.archive.krisis.eu/biocapitalism*.
56. See Foster, 'Extractivism in the Anthropocene', op. cit. (note 59).
57. See Achille Mbembe, *Critique de la raison nègre* (Paris: La Découverte, 2013), 67–68, quoted in S. Mezzadra and B. Neilson, 'On the Multiple Frontiers of Extraction: Excavating Contemporary Capitalism', *Cultural Studies* (2017) vol. 31, no. 2–3, 185–204. Online. Available at *www.datafarms.org/wp-content/uploads/2023/02/Mezzadra-Neilson-Extraction-2017-Accepted-Version.pdf* (accessed 28 June 2023). Translation by the authors.

58. Ibid.
59. Deleuze and Guattari, *A Thousand Plateaus*, op. cit. (note 1), 66.
60. Ibid.
61. See T. Morton, *Humankind: Solidarity with Nonhuman People* (London: Verso, 2017).
62. F. Türetken, 'On the Most Powerful Catalyst on the Planet', Goldsmiths Library, 2020. See also further articles, such as one in which digital technology also indicates its use for political, economic and social domination of another nation or territory: Michael Kwet, 'Digtial Colonialism: The Evolution of US Empire', *Transnational Institute* (March 2021). Online. Available at *www.longreads.tni.org/digital-colonialism-the-evolution-of-us-empire* (accessed 28 June 2023).
63. The connection between labour, nonhumanism, and metallic bodies and landscapes is apparent. The paradigmatic shift from industrial to service to data society is accelerated by REMs, which provide the necessary chemical and mechanical functions. It is within this context that elemental arteries are formed and combined with the artificial intelligence needed by the 'metallic head' to lead or reign over new formations of metallic bodies and synthetic creatures. Shudu has more than 150,000 followers; thinking of Dr Frankenstein's monster and Shudu as one and the same category of creatures, one can argue that they are both dreamt up by the ghosts of colonial phantasma. Both depend on electricity, circuits, and conductivity and data centres – in fact, it is through metal that these colonial fantasies rise again, now as digital commodities that are circulated.
64. As Fiona Hovenden claims: The cyborg can be reclaimed again, and again, from patriarchal image-making. It can keep a foot in silicon and a foot in carbon; it can run on blood and electricity. It can walk any street in the hope that it will be protected by its ambiguity. It may be wrong, and the risks are great, but it is an agent for fusing embodied, situated knowledge, and powerful fantasy. I'd like to think that Shudu is the late arrival, the partner demanded by Frankenstein; one of his own kind. The new generation of metallic creatures has arrived. Successfully animated, they walk towards us in Graphic Interchange Formats (GIFs) and emancipate themselves from their creators; leading metallic computer-generated image (CGI) lives that complement and mingle with humans' lives, they populate screens made of black liquid crystals. Soon they will be equipped with more than just their precious metals; they will also acquire affect, one of the aspects that Frankenstein's figure missed. Once the metallic androids are equipped with love and other affects, once they are integrated into society as Alexa, Siri and Shudu are, they will finally create their own kin. F. Hovenden, 'Introduction to Part Four: Refractions (women, technology and cyborgs)', in: G. Kirkup et al. (ed.), *The Gendered Cyborg* (London: Routledge, 2000), 260.
65. Superconductivity is a phenomenon in which the electrical resistance occurs in certain materials and suddenly drops to zero at its transition and cold temperatures. Substances that have become superconductive are known as superconductors; when coiled as superconducting magnets. The theory of superconductivity was established in 1957 by John Bardeen, Leon Cooper and John Robert Schrieffer. The superconductivity phenomenon was discovered in 1911 by Dutch physicist Heike Kamerlingh Onnes.
66. Abraham, *The Elements of Power*, op. cit. (note 34), 122.
67. Ibid.
68. Ibid.
69. Apart from its use in metallically extending life in cryonics, neodymium is also used to determine the age of rocks and meteorites; in electric motors and generators, mobile phones, colour television, audio systems, computer hard disks, glass used for fibre optics, astronomical work and as neodymium supermagnets.
70. R. Negarestani, *Cyclonopedia, Complicity with Anonymous Material* (Melbourne: re.press, 2008), 29; Abraham, *The Elements of Power*, op. cit. (note 34), 124.
71. This reminds me of Kiruna, the iron ore town in Sweden, that became so large that the city above the mine had to be moved. L. Lund Hebbelstrup, 'Greenland election could see halt to rare-earth mining', *euobserver*, 9 April 2021. Online. Available at *www.euobserver.com/world/151487* (accessed 25 February 2023).
72. I refer to this in 'On the Most Powerful Catalyst on the Planet', op. cit. (note 66).

Aesthetic Energies – Artistic and Cultural Perceptions of Data Centres

'Aesthetic Energies' focusses on the artistic and cultural perceptions of data centres. Data centres are not only technological and industrial facilities serving global digital economies and networks, they are also cultural artefacts. As physical and architectural structures, they occupy space and enter into largely invisible relationships with their rural or urban surroundings. They are also objects that can be speculated and theorized about from a variety of perspectives – part of narratives and histories both large and small. This chapter concentrates on the effects that the data centre as a cultural and aesthetic object unleashes within the bodies of users and its surroundings. How does it reshape the structures of feeling that organize the social world? What does it bring about in its environment?

The very nature of data centres makes it difficult to penetrate their content, which is why they are often compared to black boxes. By analysing the workings of the black box from a 'pataphysical' perspective – in the vein of late nineteenth-century Alfred Jarry and his science of imaginary solutions – such as Caleb Larsen's Internet-connected artwork *A Tool to Deceive and Slaughter*, media theorist and curator **Annet Dekker** explores the socio-technical complications and ethical implications of the often invisible elements and numerous actors in and around data centres. She argues that data centre aesthetics is located in the encounters, arrangements and combinations through which it is constructed and generates information, or in the fragments and layers that may soon be excavated.

Jorinde Seijdel's 'Data Centre Assemblage' is a collection of heterogeneous, more or less ready-made text fragments from different origins – in that sense somewhat comparable to a ChatGPT text. The glossary contains terms and references that do not serve the data centre, but rather form an affective and critical network around it. Its entries – from acid clouds to technodiversity and net-zero data centres – relate to each other but at the same are timelines outward, connecting to alternative voices and discourses.

Urbanists **Michiel van Iersel** and **Mark Minkjan** of the platform Failed Architecture evaluate the urban and social history of data centres from a futuristic perspective to find out how they are reshaping the cities and communities we live in. Van Iersel and Minkjan boldly envision the decentralization of the Internet and the re-politicization of data. They foresee the appearance of Data Community Centres 'as proud cathedrals of the digital age, collectively conceived and constructed over multiple generations'. The illustrations accompanying their text are by architect Anne Dessing, who investigates architecture through exhibitions, installations, drawings, models, interiors and buildings.

A Pataphysics of Data Centre Aesthetics

Annet Dekker

Prelude: Data Centre Aesthetics

In 2009 I encountered a square shiny black box with an ethernet cable running from its base into the pedestal on which it was displayed. *A Tool to Deceive and Slaughter* (2009) by Caleb Larsen was both a conceptual work and a physical sculpture that was wired to the vast space of the web where it was being auctioned on eBay as an 'eternal transactional flux'. This meant that it checked its eBay auction page every 10 minutes to see if any bids had been made. Once the auction deadline expired it automatically created a new auction on eBay. If a bid was successful, the owner had to send the box to the subsequent owner in line with the terms of contract. So, while the bidding and the transactions continued online, the box travelled the world. With the box supposedly becoming battered, bruised and scratched while in transit, the online transactions left few visible or directly traceable fingerprints, and the only proof of its processual state was its increasing value on eBay. The 8-inch square box became the encapsulation of autonomously moneymaking machines.

Obscuring the internal workings and emphasizing its conceptual worth, the installation referenced its predecessors in abstract art such as *The Black Square* (1915) by Kazimir Malevich, as well as in conceptual art like Robert Morris's *Box with the Sound of Its Own Making* (1961), a box with a tape recorder that played the sounds of Morris making the box; and *Money* (1969–1973) in which Morris

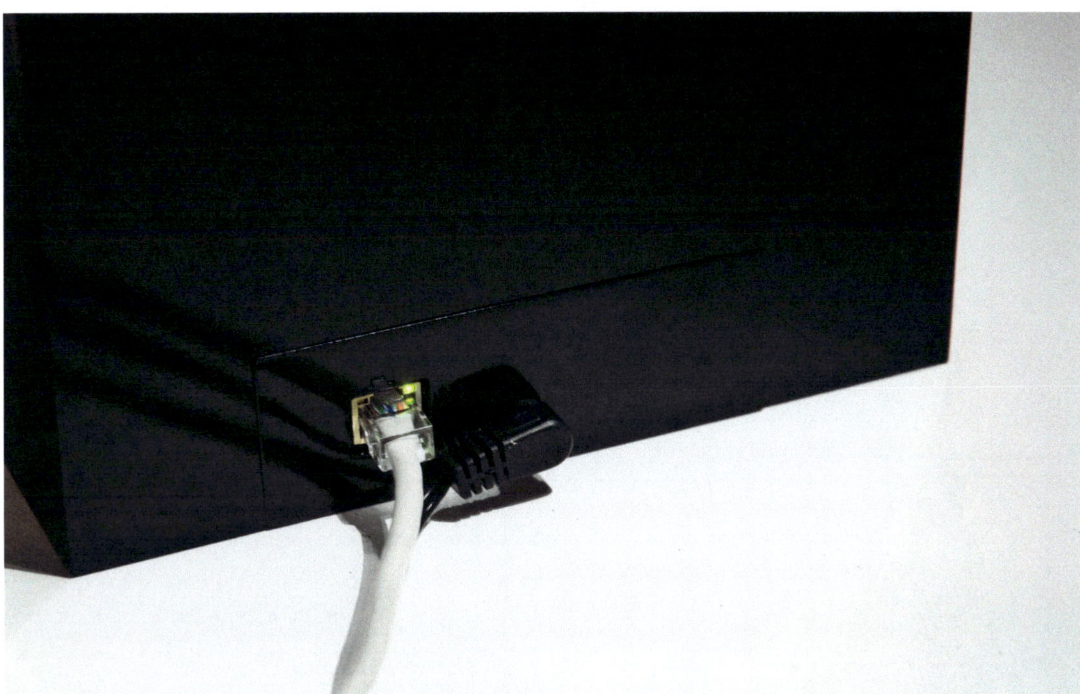

Caleb Larsen, *A Tool to Deceive and Slaughter* (2009), installation view.
Photos by Caleb Larsen

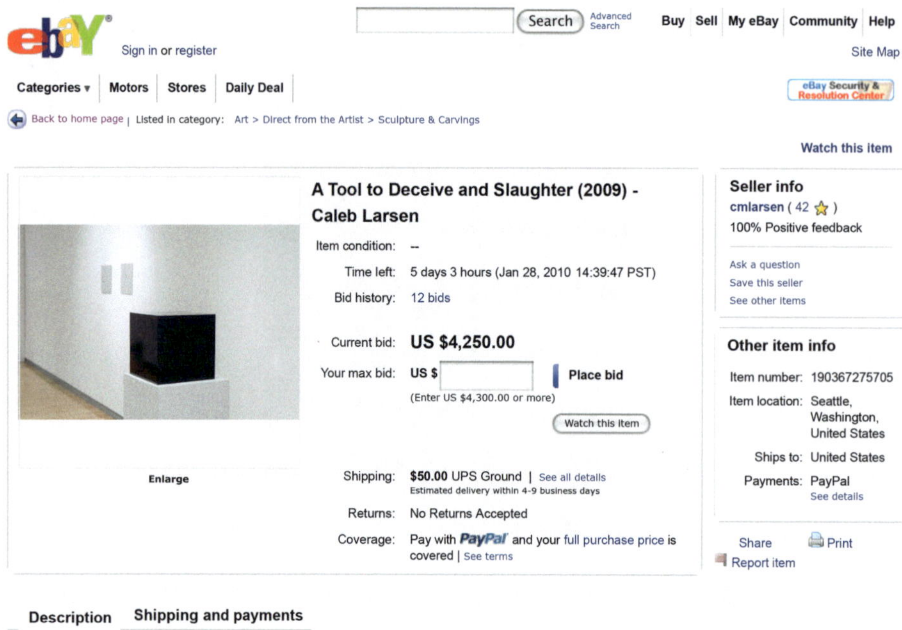

Caleb Larsen, *A Tool to Deceive and Slaughter*, eBay auction site (screenshot, 23 January 2010, s3.amazonaws.com/giles/tool_012310/A_Tool_to_Deceive_and_Slaughter.pdf).

asked the Whitney Museum Museum of American Art in New York, amid a monetary crisis, to invest $100,000 in the stock market; or the two black monoliths in Stanley Kubrick's *2001: A Space Odyssey* (1968) that signify the machines built by extra terrestrial species to change the course of life on Earth. Both mimicking the form (black box) and conceptual content (the artist's anti-market attitude as well as its subsequent capitalization and the reference to ominous futures), Larsen linked the immateriality of financial art trading to the current 24/7 society with its emphasis on speed, enduring interconnections and a continuous stream of data that is coupled to an insatiable appetite for information as well as storage and retrieval.

The black box has been a prolific metaphor for all that needs to be concealed. During World War II the term became popular when it was used to name the physical black-boxed systems or objects that produced useful information but did not reveal their inner workings to the outside world. While Larsen's *A Tool to Deceive and Slaughter* is a humorous nod to the sinister inside black boxes, the aesthetics of the 'black box' has shifted from classified information devices to all other kinds of relays; one of them being physical data centres. Whereas most data centres remain similarly opaque, what are the aesthetic registers of modern-day data centres? Besides its shape, a data centre is involved in many other relationships: from its physical location connecting to its virtual cloud; from the

people who traverse obscured doors to those who observe the black box from their neighbouring windows; from the thousands of invisible 'Mechanical Turks' that work on raw data to those who upload, read and (re)use information; or from the electricity that powers the servers to the cables that distribute the data and connect to a multitude of other data centres. [1] What would be the value of data centre aesthetics? What new knowledge about, or perhaps ethics of, data centres could such an expanded view provoke?

Larsen sees his project as 'site-less', meaning that it is able to cross the boundaries of physical and virtual space. Being part of an information space that is governed by sets of rules, or protocols, the black box symbolizes the construction and manipulation by programmatic systems. [2] In a conventional sense this means that the aesthetics of the project is valued as a combination of the concept (who created it, or the ideas of the artist), form (what it looks like, or the physical materiality) and content (what happens in or with the project, or the actions of users). However, data centre aesthetics attempt to develop an understanding of a wider set of complexities and processes, for instance, around time and ownership, that are intricately connected with various processes: from land-use planning, to a global network of power cables, to ecological damage as dynamic processes of composition, construction and power. Working within the framework of data centre aesthetics, and following Larsen's approach, I will focus on the data centre as operating in the 'Pataphysics': 'the science of imaginary solutions, which symbolically attributes the properties of objects, described by their virtuality, to their lineaments.' 3 Perhaps this is an optimistic and absurd attempt to navigate alternate paths within the economic and political structures of profit-orientated data centres. However, by exploring the socio-technical entanglements and hence the ethical implications of the often-invisible elements of and the numerous actors in and around data centres, I want to emphasize the 'transactional value' of data centres. 'Transactional' not necessarily in the sense of monetary exchange, but rather in the relationship between the material, its construction and functioning, including the different actors involved in and around the data centre. Hence, I'm interested in identifying the moment when the value of a data centre becomes the essence of its functioning.

Start from the Beginning: Time

Larsen asked his potential buyers to adhere to his Purchase Agreement. The stringent rules consisted of 18 conditions, the first being that the cube should 'remain connected to a live Internet connection at all times, with disconnections allowed only for transport of the work'. 4 Being connected 24/7 is often seen as the systemic imperative of twenty-first century capitalism. 5 Time and technology are intricately entwined: from the way time is measured and translated to 'clock-time', also known as 'clock ticks' or 'cycles', 6 to how technology is seen as a tool that saves time and improves efficiency and productivity. Sociologist Georg Simmel described how the watch became crucial in defining the notion of time: the precision of the quantification of time accelerated the structuring of industrialization. 7 Similarly, communication scientist James W. Carey emphasized how the invention of global standard time zones were a direct consequence of the introduction of the telegraph, which impacted the speed at which information was sent. 8 Data centres have been around since the first computers were assembled, yet the specific term emerged in the 1980s and gained currency in the 1990s when its characteristic black-box architecture was introduced. Driven by the need for rapid computational transactions and speedy information delivery, the urgency of transmission time has become the trope of many data-centre discussions that focus on the importance of colocation. Indeed, a data centre is all about site-specificity in order to reduce latency so as to achieve the best high-frequency financial trading, to execute high-precision warfare, to watch streaming video, to debate in video conferencing and to win multiplayer games.

What has received little attention is what time means inside the data centre, and more precisely inside the computer: what does time mean in a computational sense? Every nanosecond data is being crunched through wires to reach potential users/usages – this is not to say that all data is constantly processed; mostly, data is just data, merely lingering and waiting for something to happen to it. Yet, this situation led to the proposition of the detemporality of time: doing away with the idea of time as process in favour of the acceleration of information as a temporal compression. 9 In other words, where before there was a sense of events unfolding over time, due to the speed of transactions there is now almost instant delivery. Such observations tend to see the web as fluid and continuous, as a pleasant, uninterrupted flow of data and exchanges. However, in most experiences the web is more unpredictable, less like a steady stream and more like a wild river moving at different speeds, with downtimes, blockages, missing parts, relays and dead ends. In that sense, rather than detemporality, it is closer to a temporal discontinuity, or broken time. 10 When analysing the working of the Viterbi algorithm, 11 sociologist Adrian Mackenzie listed the various types of machine time: seek time, run time, read time, access time, real time, polynomial time, time division, time slicing, time sharing, time complexity, processor time, execution time, compilation time ... Emphasizing this complexity Mackenzie explains how the Viterbi algorithm shows that its future is only partially dependent on the past, as it is based on a technique where the future outcome is traced to potential past causes of the current event. He continues: 'The incorporation of past states into each message begins to break down the punctual, discrete nature of repetition. A flow of information begins to look more like a phenomenological "now", composed of retentions and protentions.' 12 So, whereas watches have become an accepted way to relate to time, computing-based time is less straightforward and requires a more agile comprehension of various temporalities that influence each other. Indeed, in the complex mix of past memories and future predictions, the present can be hard to pin down inside the data centre.

Could a similar sense of timelessness be observed in data centre employees? What would a human experience of time mean in data centre aesthetics? Jobs in data centres range from technicians installing and maintaining network resources, ensuring data centre security, monitoring systems that take care of power and cooling, to cleaners mopping the empty corridors and dusting the server racks, and helpdesk employees or service delivery managers addressing questions from present or potential clients. According to recruitment ads, time is not an issue: 'What I do is definitely not a nine-to-five job. I maintain the servers, repair them, diagnose issues. It involves a lot of troubleshooting and lots of physical work. You're never bored here. You have to be on top of your game, every minute counts.' 13 Time is even less tangible for workers, aka Turkers, who work with the data that data centres rely on. 14 Despite the often-used slogan that new data centres will bring an economic boost by creating more jobs in the local environment, in reality a typical data centre employs between 5 to 30 people. 15 Clearly this doesn't take into account the armies of people behind the scenes who try to make sense of data and steer the functioning of algorithms: those who look at, annotate, classify and submit captions within milliseconds. While the long-term investment in labour employment is not necessarily high, local and national government bodies calculate their revenue based on the short-term investment. This relates to the construction of the black box, the purchase of computer servers, the use of electricity and the subsequent payment of taxes. By choosing to profit in the present rather than in the future, a human experience of time in relation to data centres is reflected in the imbalance of time: favouring the short-term (the present) over the long-term (the future). Whereas the computational transactions inside the data centre are timeless, the human construction and execution of the data is focussed on the present.

Data centre time is a complex assemblage in which computer-based times and the traces of human intervention become entangled. This merge generates potentially unlimited experiences of temporality without a clear trajectory, oscillating between timelessness and presentism. Yet, with a better understanding of the increasing dependencies between the machines and the human labour force, data centre aesthetics can help to develop ways of understanding that provoke questions and critique, even if implicitly, through the actions a data centre engenders.

Midway: Ownership

Another remarkable condition in Larsen's Purchase Agreement is that each owner of *A Tool to Deceive and Slaughter* can assign a new value to the black cube, abiding by 'current market expectations for the Artwork based on the current value of work by the Artist'. Yet, immediately after being purchased *A Tool to Deceive and Slaughter* is relisted on eBay, and the new collector is 'required to sell the Artwork to highest bidder'.[16] Regardless of whether it is a private collector or a museum, no one can own the project in perpetuity. Ownership merely lasts as long as the auction is live, or as mentioned by Larsen, 'ownership is sited in the moment of transaction'.[17] As an uncollectible artwork, *A Tool to Deceive and Slaughter* undermines ownership, and while it generates a tension about the role of collectors in determining the value of an artwork, the market is now open to any collector – if they have enough funds to bid on eBay – and is not dependent on the artist's preference. *A Tool to Deceive and Slaughter* reflects financial trading, where, as a consequence of virtual assets that stimulate the trading of derivatives on derivatives, what one actually owns is becoming increasingly opaque.

In the past, most companies built their own storage facilities, so-called 'onsite data centres', where they could compute, store and network their data. However, the economics around data centres has shifted in the past few years. Initially the

The cover of *Time* magazine's 2006 Person of the Year issue was dedicated to 'You' emphasizing the explosive growth and influence of user-generated web content such as blogs, the video-sharing platform YouTube and the social network MySpace. 22 Photo-illustration for *Time* by Arthur Hochstein, with photographs by Spencer Jones-Glasshouse.

focus was on building one's own physical infrastructure, where the data was easily accessible and could be quickly managed or maintained when required to keep the data secure and moreover, was controlled by the company itself. Yet, under the dictum of efficiency and competition there are several disadvantages as data centres tend to grow and consequently consume an increasing amount of resources: from a reliable power supply and cooling systems to a large network and an extensive security system. Hence, many businesses rent space in a colocation facility, also known as a 'colo', or use the service of a hyperscale data centre. Whereas in the former, companies could still own their data servers, in the latter these are leased from the data centre. In return one is given server space with guaranteed up-to-date maintenance and security, as well as a stable Internet connection. Customers started to upload their assets into the 'public cloud', primarily for easy access at anytime, anywhere and on any device.[18] The shift towards leasing space also impacted the thinking about ownership. The traditional long-term thinking about assets that would be self-owned and easily accessible shifted to short-term objectives that favoured efficiency and operational flexibility while accepting the loss of the certainty of ownership and control. Indeed, what does ownership mean in the cloud?

Ownership refers to the legal rights of individuals, groups, corporations or governments to possess something, either material and immaterial. Material ownership relates to things that are tangible: property, land, cars, books, computers, etc. Immaterial ownership applies to things that are intangible: such as patents, copyrights and trademarks. The distinction between material and immaterial has been widely debated in relation to the Internet, and digital data more generally. As stated by Marianne van den Boomen, 'the productivity of the shiny screen of representation and reification pushes itself to the fore, [and the] unacknowledged black box is working hard in the background to accomplish numerous invisible transferences'.[19] Similar to a data centre, Larsen's black box conceals its material inside and only presents its immaterial outcome. Although most people realize now that the working of the computer has many material properties and consequences, the prevalent metaphor of the invisible and immaterial is resurrected with 'the cloud'.[20] However, this time the discussion seems to be less about how the cloud works and more about who owns it, and more pertinent, who owns the data in the cloud. Whereas ownership is suggested in many advertisements for computer programs that 'shamelessly use shifters, pronouns like "my" and "you," that address you, and everyone else, as a subject',[21] the ownership over 'my' data is more elusive.

Cloud providers store data, and usually several copies for reliability; in keeping with the notion of LOCKS (Lots Of Copies Keep Things Safe), there is always a backup in the event a server fails. Yet, details about how many copies are made and where they are stored are sparse. This perhaps insignificant detail becomes more important when relating your data to specific laws and regulations. For example, in the EU data that is held is subject to the EU Data Protection Directive – and there may be consequences when compared to other nations, as the case of Edward Snowden demonstrated. 23 Although data providers are obliged to protect your data from misuse, data breaches are far from uncommon. Similarly, although there is legislation in place that empowers users to ask providers (or other organizations) to delete their data, this Right To Be Forgotten (or right to erasure) as part of the European GDPR (General Data Protection Regulation), is not always applicable. 24 Moreover, there is often a (undetermined) delay in the process of deleting data. Most regulators insist that the providers are responsible for the security of the cloud, while the customers are responsible for the security in the cloud. 25 In other words, securing personal data is the owner's responsibility. So, back to ownership... With several copies of your data stored at various data centres in different locations, the labyrinth of legalese may close in around you. However, this is solved by the development of 'safe data harbours': an environment that consists of data-centric security technologies and offers cloud scalability with precision control. Here data is automatically fragmented and scattered across multiple storage locations, still in the cloud but now more easily traceable. This procedure prevents an unwanted attack onto the data, and since it no longer exists in a usable (original) state, it circumvents any jurisdiction. 26 But what happens to the now-fragmented data?

Streams of data are continuously generated in the cloud: client data, data from third-party service providers, application data, process data, and more, are compiled, accessed, parsed, manipulated, combined, transmitted, stored, re-transmitted and reused in cycles that can repeat endlessly. While these processes can be divided in two types of data streams – customer (user) data and the data centre data that is needed to store all the data, which includes security encryption measures, and other applications by service providers – finding a way through the maze of data traffic requires expert knowledge. Despite many attempts to create maps, guides and visualizations of the (inner) working of clouds or data centres, and although some may have this expertise, within the inherently intra-networked infrastructure of data centres, complexity rules and it is increasingly difficult to determine a single data file, let alone a fragment of a file. Hence, the question of ownership is not so much about who owns data, but more about how to access and reuse it? In other words, who controls data, and who (as a user) do you trust, or (as a service provider) how to be trustworthy? Ownership implies – even if it's only at the moment of transaction – issues of power and consequently it concerns control, trust and ethics. To some the ambiguity of the black box, the transactions that are made and its wavering ownership or the loss of control are exactly what attracts them, as mentioned by Terence Spies, one of the buyers of *A Tool to Deceive and Slaughter*, when asked why he was interested in the piece: 'I had a really strong reaction right after I won the auction. I have this thing, and I really want to keep it, but the reason I want to keep it is that it might leave. ... [it] really exposes your motivations for wanting to own something.' 27 In the comment section of the article, someone remarks: 'The thing I *love* the most about this work being on eBay is that there is the big "eBay Buyer Protection" banner telling me I'm protected from loss."' 28 Indeed, what happens to your data when you give it up and it gets lost either in transaction or in the (near) future?

Ending: Torpor, or the Eternal Transactional Flux Doesn't Last

The last auction of *A Tool to Deceive and Slaughter* ended on 19 April 2017 when it sold for US $7,500; at least, this was the last bid I found. This sliver of information was no longer on eBay, or on Larsen's website, both of which ceased to exist some time ago. The only fragmented piece of evidence was lying dormant in the vault of the Wayback Machine. In a post on Twitter, on April Fool's Day 2014, Larsen stated the project was still ongoing, even though the box had been damaged in the process. Was the posting on 1 April a coincidence? Did the box actually ever travel anywhere?

From the interview with Spies, it seems that he was planning to go to the exhibition where *A Tool to Deceive and Slaughter* was presented to see the black box before it was sold to the next highest bidder. The notion of 'deceive' in the title of the project could be interpreted in various ways: in the sense of a misleading notion of ownership; as a critique of the art market; as a critical nod to the mystical symbolism of the black box in art practices, religion or science fiction, as well as in connection to the increasingly inaccessible inside of data centres and computers more generally; or as a prank on each of these – including those trying to comprehend its (inner) workings, i.e., yours truly, all to be slaughtered in the process?

In the meantime, *A Tool to Deceive and Slaughter* has moved from an eternal transactional flux into torpor. The project has come to resemble the amounts of data in the cloud that are shunted about by electrical currents but remain unused. Some estimate that around 80 per cent of companies' data in the cloud is dormant: also known as dark data, it is collected and stored data that is not analysed or used to inform decisions or make predictions. While it has become impossible to estimate or even approximate the exact amount of data in the cloud, it is clear that only a nominal portion is actively deployed. This quantity could be said to be irrelevant, as the nature of the apparatus is to condense the volume of data, to extract that which is 'probable or possible, enacting what matters and what is excluded from mattering'. 29 Similarly unknown, and perhaps

Caleb Larsen
@cmlarsen

Follow

After touring the world & getting damaged, my eBay box, A Tool to Deceive and Slaughter is back online! atooltodeceiveandslaughter.com

9:06 AM - 1 Apr 2014

1 Retweet 1 Like

2 1 1

Caleb Larsen, Twitter post,
1 April 2014, screenshot.

more urgent, is the likelihood of the survival of data. In other words, the sustainability of digital data is unclear. While Larsen's black box may still be in a depot somewhere, battered and all, its active online life seems to have lasted for about eight years. The project didn't withstand the test of time. Likely due to hardware and software updates, it was 'slaughtered' by its own environment, leaving only fragments behind. The same will probably be true for data centres: besides being dormant the data will be inaccessible in a decade or so due to neglect or having ceased to endlessly upgrade hardware and software, while the architectural box continues to decay, slowly succumbing to environmental challenges. Yet, these withering remains could be seen as key to the survival and retelling, or referencing, of data.

Indeed, what would it mean if these fragments were taken as points of departure to consider the future? What does it mean when data centres do not enable coherent thinking about the accessibility of their data (either by pursuing safety harbours, unlocking dormant data or mitigating security challenges), and instead 'stay with the trouble', as Donna J. Haraway puts it,[30] of partial and indeterminate pieces of data, thereby shifting from active perpetual systems to partial ways of comprehending worlds? In other words, this means accepting that digital data cannot be considered in the same way as conventional material objects, and

hence the methods that are used to sustain it need to be adapted to reflect the inevitable change. Moreover, by acknowledging how the data evolves, and taking cues from its 'immateriality' and its volatile transactional nature, this future 'data', as well as data centres, can become useful again, albeit in different ways: fragments can be used to tell alternative stories that make sense in the present. The debris will be mixed with other rubble to create new relevance, much as I am trying to make sense of data centre aesthetics through examining remnants of *A Tool to Deceive and Slaughter*. I fabricated new knowledge: knowledge that is partly based on facts and blended with some fiction, and where even dormant data is rendered commensurate to be useful again. Digital data, and data centres, shift preservation from 'static' storage to an active site of generative (un)imagined futures: to continue in perpetuity.

Postscript

A title stimulates an associative process, however, *A Tool to Deceive and Slaughter* was nothing more than a line of text written in the lavatory of a 'grimy, smoke-filled, 24-hour coffee shop in Kalamazoo, Michigan'.[31] Without context this title has no meaning, makes no sense and is merely a bold or fun gesture that only gains value in the domain of toilet poetry. Similarly, the black box is merely a reflection on the status of the art object as a constructed entity, and likewise, the data centre represents data as raw material, which, when processed, produces information and hence insights. The quest for meaning is one of the main drivers of most Western scientists. When people asked Larsen about the inside of the black box on fora such as Reddit, he was unambiguous about how he programmed the work. In this sense, Larsen's project is not about mystification or the invisibility of the black box, but about rendering perceptible and actionable that which is usually not discussed or plainly ignored: the power of art economics and the dominance of ownership.

Shannon Mattern explains how the presumption that data centres, or in her case, infrastructures, 'are "hidden" or "magic," and thus require demystification through a field trip or field guide, signals great privilege'. Indeed, and as she continues, 'all over the world, there are cable-layers and data-centre operators, cell-phone salesmen and e-waste handlers who have long possessed trade-specific knowledge about the Cloud's presence and operations – knowledge that we've only recently come to valorise'. 32 Attempts to visualize, demarcate or quantify the workings of data centres are abundant, yet often they generate merely partial views. My brief quest for comprehensibility is also measured against specific beliefs and tends to favour certain contraptions and views over others, comparable to appeals for transparency, standardization or accountability. To say that the black box somehow obfuscates a real world of geopolitics could likewise be seen as upholding the mechanism that one tries to address. Instead, in analogy to Mattern, what would happen if the tropes of mysterious black boxes, big data and algorithmic coherence are rejected, and the question of 'where or what is in the data centre?' is replaced with 'how does a data centre perceive and analyse the world?'.

According to Larsen the black box is merely a tool to deceive and slaughter. To understand its inner workings is unfeasible – indeed, many programmers affirm that even they don't know why and how their programs function. Moreover, it is not something to desire. It's more interesting to mess around in the unknown and to see the object or space for what it is: fertile ground where one can linger, hesitate and doubt, as meaningful moments of socio-ethical assessment in an attempt to reimagine the attributes of its 'black box'. This is not about not taking responsibility, but about accepting that all knowledge is always partial and contextual, and thus, about embracing the inherent anomalies of the complex actions that take place in human and non-human interactions, and how these account for new patterns that emerge with each new input. Hence data centre aesthetics is located in the encounters, arrangements and combinations through which it is constructed and generates information, or in the fragments and layers that may soon be excavated. It is about questioning the authorial subject or source and opening up the multiple forms of creating and distributing data, in the process leaving ownership uncertain and its emergence opaque.

1. The term 'Mechanical Turk' originates from a chess-playing machine that was developed in the late nineteenth century by the Hungarian inventor Wolfgang von Kempelen. Promoted as an automaton, in reality a human being hidden inside the chess cabinet made the chess moves. The term gained new relevance with Amazon Mechanical Turk (2005), an online marketplace (www.mturk.com) for jobs that could more easily be done by humans than machines and which later was used more generally to describe the (invisible and low-cost) human labour involved in 'automated' digital processes. For more information about the Mechanical Turk, see Pew Research Center, 'Research in a Crowdsourcing Age', 11 July 2016. Online. Available at *www.pewresearch.org*/internet/2016/07/11/research-in-the-crowdsourcing-age-a-case-study (accessed 28 June 2023), and about the labour conditions and ethics of MTurks, see, for instance, M.L. Gray and S. Suri, *Ghost Work: How to Stop Silicon Valley from Building a New Global Underclass* (Boston: Houghton Miffin Harcourt, 2019).
2. C. Larsen, *The Value of Nothing* (exh.cat.) (Providence: Rhode Island School of Design, 2009), 17.
3. A. Jarry, 'What is Pataphysics?', *Evergreen Review* (1963) vol.4, no.13, 131. Most definitions of Pataphysics see it as a branch of philosophy or science that examines imaginary phenomena beyond metaphysics. As described by Georges Perec, 'If physics proposes: "You have a brother and he likes cheese", then metaphysics replies: "if you have a brother, he likes cheese". But Pataphysics says: "You don't have a brother and he likes cheese".' 'What Is Pataphysics', *Musée Pataméchanique*, no date. Online. Available at *www.museepata.org*/what-is-pataphysics (accessed 28 June 2023). In other words, Pataphysics uses the method of divergent thinking.
4. See Larsen, *The Value of Nothing*, op.cit. (note 2).
5. See J. Crary, *24/7: Late Capitalism and the Ends of Sleep* (Brooklyn, NY: Verso Books, 2013).
6. For more information, see 'Computer Clock – Clock Tick (Clock Cycle)'. Online. Available at *www.datacadamia.com*/computer/clock/tick (accessed 28 June 2023).
7. See G. Simmel, 'The Metropolis and Mental Life', in: G. Bridge and S. Watson (ed.), *The Blackwell City Reader* (Oxford and Malden, MA: Wiley-Blackwell, 2002 [1903]), 11–19.
8. See J.W. Carey, 'Technology and Ideology: The Case of the Telegraph', *Prospects* (October 1983) vol.8, 303–325.
9. M. Castells, *The Rise of Network Society. The Information Age. Economy, Society, and Culture, Vol.1*, 2nd ed. (Oxford: Wiley-Blackwell: 2010), 464.
10. See A. Dekker, 'Curating Digital Art: A Synthesis of Space and Time', *Planetary Atoll: Connecting Latin American Dots* (exh.cat.) (Berlin: panke.gallery, 2023).
11. The Viterbi algorithm is a dynamic programming algorithm, which is used to compute the most probable path, called the Viterbi path. It works by asking a question like 'given the trained parameter matrices and data, what is the choice of states such that the joint probability reaches maximum?'. See also, R. Kwok, 'Viterbi algorithm for prediction with HMM – Part 3 of the HMM series', *Medium*, 21 July 2019. Online. Available at *www.medium.com*/analytics-vidhya/viterbi-algorithm-for-prediction-with-hmm-part-3-of-the-hmm-series-6466ce2f5dc6 (accessed 28 June 2023).
12. A. Mackenzie, 'Protocols and the Irreducible Traces of Embodiment: The Viterbi Algorithm and the Mosaic of Machine Time', in: R. Hassan and R.E. Purser, *24/7: Time and Temporality in the Network Society* (Stanford, CA: Stanford University Press, 2007), 104.
13. Lou interviewed by About Amazon Team, 'What's It Like to Work in Data Centers?', *Amazon News*, 10 March 2020. Online. Available at *www.aboutamazon.eu*/news/working-at-amazon/whats-it-like-to-work-in-data-centers (accessed 28 June 2023).
14. See N. Malevé, 'The Computer Vision Lab: The Epistemic Configuration of Machine Vision', in: A. Dewdney and K. Sluis, *The Networked Image in Post-Digital Culture* (London: Routledge, 2023), 83–101.
15. See J. Lenio, 'The Mystery Impact of Data Centres on Local Economies Revealed', *AreaDevelopment* (2015). Online. Available at *www.areadevelopment.com*/data-centers/data-centers-q1-2015/impact-of-data-center-development-locally-2262766.shtml (accessed 28 June 2023).
16. See Larsen, *The Value of Nothing*, op.cit. (note 2).
17. See ibid.
18. The distinction between a data centre and cloud storage seems insignificant, yet the difference is that a data centre stores data on the premises of either an organization or collocation. This doesn't always mean that the data is also available in the cloud, i.e., that data is accessible from anywhere via the Internet.
19. M. van den Boomen, *Transcoding the Digital: How Metaphors Matter in New Media. Theory on Demand #14* (Amsterdam: Institute of Network Cultures, 2014), 16.
20. The image of the cloud did not come out of nowhere; computer specialists commonly use a cloud cartoon in systems that are opaque. S. Frye, 'Network Diagrams Highly-rated by IT Pros', *TechRepublic*, 2 September 2010. Online. Available at *www.techrepublic.com*/pictures/network-diagrams-highly-rated-by-it-pros/ (accessed 28 June 2023).

21. W. Hui Kyong Chun, 'On Software, or the Persistence of Visual Knowledge', *Grey Room* (Winter 2004) vol. 18, 43.
22. At the same time, it also signalled the shift from 'my' to 'me' (you): the former registers possession (a possessive pronoun) whereas the latter simply refers to 'yourself' (an object pronoun). As discussed by net artist Olia Lialina when describing changes in web culture, this also means a change of ownership in which there is a shift from owning your data to giving it away. O. Lialina, *Turing Complete User. Resisting Alienation in Human-Computer-Interaction* (Heidelberg: Heidelberg University Publishing 2021), 191.
23. In June 2013, the, at the time, National Security Agency contractor Edward Snowden released a trove of information on classified US government surveillance methods. While the leak led to his own loss of freedom, it also spurred the development of the European Union's General Data Protection Regulation (GDPR) that changed the nature of data privacy in the EU.
24. For the latest updates, see 'Complete Guide to GDPR Compliance', GDPR.EU. Online. Available at *www.gdpr.eu* (accessed 28 June 2023).
25. See, for instance, 'Shared Responsibility in the Cloud', Microsoft, 12 June 2022. Online. Available at *learn.microsoft.com/en-us/azure/security/fundamentals/shared-responsibility* and Amazon, *aws.amazon.com/compliance/shared-responsibility-model* (accessed 28 June 2023).
26. P. Lewis, 'Can Data Be "Jurisdiction-independent"?', *Calamu*, 30 April 2021. Online. Available at *www.calamu.com/blog/jurisdiction-independent* (accessed 28 June 2023).
27. J. Graves, 'Buying Art That Just Wants to Escape From You: A Conversation With the Collector of a Tool to Deceive and Slaughter', *The Stranger*, 2 February 2010. Online. Available at *www.thestranger.com/visual-art/2010/02/02/3343186/buying-art-that-just-wants-to-escape-from-you-a-conversation-with-the-collector-of-a-tool-to-deceive-and-slaughter* (accessed 28 June 2023).
28. Ibid.
29. L. Amoore, *Cloud Ethics. Algorithms and the Attributes of Ourselves and Others* (Durham, NC: Duke University Press, 2020), 14.
30. See D.J. Haraway, *Staying With the Trouble: Making Kin in the Chthulucene* (Durham, NC: Duke University Press, 2016).
31. See Larsen, *The Value of Nothing*, op. cit. (note 2), 31.
32. S. Mattern, 'Clouds and Fields', *Places* (August 2016). Online. Available at *www.placesjournal.org/article/cloud-and-field/?cn-reloaded=1&cn-reloaded=1#footnote_79* (accessed 28 June 2023).

Data Centre Assemblage: Is There Something Else You Would Like to Know?

Jorinde Seijdel

This glossary attempts to understand the data centre as part of social and (im)material networks and assemblages of human and non-human actors, effects and affects. It consists of bits and pieces – quotes, paraphrases, descriptions, concepts, links, references – of various, sometimes questionable origins and sources, evoking automated and / or dissenting voices. The entries may interact and thus reinforce each other, prompting larger reflections that are concerned with challenging the notion of objective data storage and dissemination and resisting dominant power relations and (in)equality within digital capitalism. The illustrations accompanying the glossary were drawn by Alice Mulder.

Acid Clouds

Acid clouds twist, data's ballet,
Silent hums merge, in arrays.
Bytes once vibrant, now corrode,
Tech's promise tarnished, we erode. 1
– ChatGPT

Ada Lovelace (they/them)

First: 'Augusta Ada King, Countess of Lovelace (*née* Byron; 1815–1852) was an English mathematician and writer, chiefly known for her work on Charles Babbage's proposed mechanical general-purpose computer, the Analytical Engine. She was the first to recognise that the machine had applications beyond pure calculation.' The early programming language Ada was named after her, and the second Tuesday in October has become Ada Lovelace Day, on which the contributions of women to science, technology, engineering and mathematics are honoured. 2

Second: Ada Lovelace is the codename for a graphics processing unit (GPU) microarchitecture developed by Nvidia, officially announced 20 September 2022. Nvidia's GPU architectures, including Ada Lovelace, are relevant to data centres due to their powerful computing capabilities and efficiency in handling various workloads, particularly those related to artificial intelligence, machine learning, high-performance computing and data analytics. 3

Ai-Da

Ai-Da is a humanoid robot artist that came into being in 2019. Ai-Da's name is a combination of the acronym for artificial intelligence (AI) and a reference to Ada Lovelace.

'In the time of increasing use of AI, online avatars, AI chatbots, Alexa and Siri, – Ai-Da as a robotic artist is acutely relevant. She is not alive, but she is a persona that we relate and respond to. This surreal situation of confusing realities is already part of our daily lives: in our digital realms, who are we speaking to on online platforms? What algorithms are working behind our internet choices? Who writes the algorithms, and who benefits and who loses? Extraordinarily complex, our online worlds are pushed and pulled by forces and personalities that are sometimes apparent but largely oblique. Ai-Da, the machine with AI capacities, highlights those tensions: bringing to the forefront the complexity of our interacting digital and physical worlds and the masked identities we can assume in them.' 4

AI for the people!!!!! 5

Algorithm Observatory
→ *see Data Colonialism*

A collaborative online lab for the study of social computing algorithms. The principal investigator is Ulises A. Mejias, co-author with Nick Couldry of *The Costs of Connection: How Data Is Colonizing Human Life and Appropriating It for Capitalism*. 6

'The Algorithm Observatory platform will allow researchers to design an experiment by focusing on a specific algorithm (a search engine algorithm, or a news recommendation algorithm, for instance), and specifying how subjects are to interact with it (submit a specific query, or access it at a particular time, etc.). AO will allow researchers to recruit participants and collect information. Researchers will then conduct an analysis to compare how the algorithm handles users differently depending on individual characteristics. The results will be shared by publishing a report evaluating the social impact of the algorithm. All data and reports become publicly available and open to comments and reviews. Any algorithm can be studied, because the site does not require direct access to the source code (which is usually proprietary) but relies instead on empirical observation of the interaction between the algorithm and volunteer participants.' 7

Algorithmic Anxiety

Patricia de Vries addresses the artistic engagement with algorithms that thematize the dissemination and deployment of algorithms in everyday life. Especially striking, says De Vries, is the 'high volume of artistic engagements with facial recognition algorithms, trading algorithms and search engine algorithms'. The fact that these 'three types of algorithms have garnered more responses than other types' suggests that they form a popular subject within artistic critique. This critique addresses 'several significant, supra-individual anxieties of our decade: socio-political uncertainty and polarisation, the global economic crisis and cycles of recession,' and the centralization and corporatization of access to online information. De Vries proposes 'that the work of Søren Kierkegaard – one of the first theorists of anxiety – can help us to investigate and to analyze' the constituents of 'algorithmic anxiety'. 8

■ Algorithms of Oppression

Safiya Umoja Noble's book *Algorithms of Oppression: How Search Engines Reinforce Racism*, 9 investigates how negative biases against women of colour are embedded in search engine results and algorithms. Noble studies how algorithmic information systems – Google Search in particular – represent people who are not white and not male. Years ago, she noticed, in shopping for her nieces, that Google responded to searches for black girls with porn assuming that's what you were looking for. 'Algorithms are, and will continue to be, loaded with power,' 10 writes Noble. 'Algorithmic oppression is not just a glitch in the system but, rather, is fundamental to the operating system of the web.' 11

■ Alienation

Jorinde Seijdel: Where is your data centre?

ChatGPT: I don't have access to real-time or location-specific information, and I don't possess awareness of my own physical location or the specific data centre that I operate from. I exist as a program running on servers within data centres managed by OpenAI, but the exact locations of these data centres are not publicly disclosed for security and operational reasons. Is there something else you would like to know? 12

■ Amnesia

ZZZ 13

■ Autonomy Cube

Sculpture by Trevor Paglen and Jacob Appelbaum, 'designed to be housed in art museums, galleries, and civic spaces. Several Internet-connected computers housed within the work create an open Wi-Fi hotspot called *Autonomy Cube* wherever it is installed. Anyone can join this network and use it to browse the Internet. But *Autonomy Cube* does not provide a normal Internet connection. The sculpture routes all of the Wi-Fi traffic over the Tor network, a global network of thousands of volunteer-run servers, relays and services designed to help anonymize data.' Paglen and Appelbaum raise the questions: 'What would a more civic-minded version of the Internet look like? What could the Internet look like if the Internet hadn't been turned into the greatest means of mass surveillance in the history of humanity?' 14

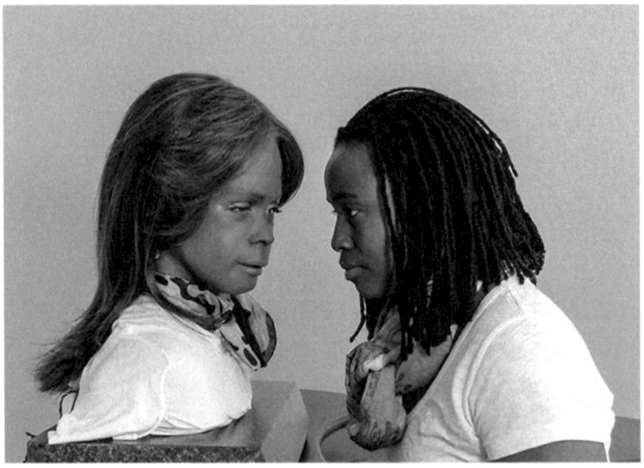

■ Bina48

'Since 2014, artist Stephanie Dinkins has been recording her interactions with Bina48, 15 an advanced social robot in the form of a plastic bust with the features of an African-American woman. Their conversations have revolved around notions of intelligence, emotions, family, faith and race. Through this process, Dinkins has been exploring the forms a human-machine relationship could take while examining the encoding of sociocultural perspectives and biases in this artificial system. Although the robot is based on the looks and memory of Bina, an African-American woman partner of millionaire

Martine Rothblatt who purchased the robot in 2010, the people who programmed it are primarily white men from Hanson Robotics. Dinkins questions if and how Bina48 could take the identity of a black woman by using her conversations both as a test and as a form to re-encode other histories into the robot.' 16

■ **Cloud (1)**

'Like the inaudible hum of the electrical grid at 60 hertz, the cloud is silent, in the background and almost unnoticeable. As a piece of information flows through the cloud – provisionally defined, a system of networks that pools computing power – it is designed to get to its destination with "five-nines" reliability, so that if one hard drive or piece of wire fails en route, another one takes its place, 99.999 percent of the time. Because of its reliability and ubiquity, the cloud is a particularly mute piece of infrastructure. It is just there, atmospheric and part of the environment.

Until something goes wrong, that is. Until a dictator throws the Internet "kill switch," or, more likely, a farmer's backhoe accidentally hits fiberoptic cable. Until state-sponsored hackers launch a wave of attacks, or, more likely, an unanticipated leap year throws off the servers, as it did on February 29, 2011. Until a small business in Virginia makes a mistake, and accidentally directs the entire Internet – yes, all of the Internet – to send its data via Virginia, and, almost unbelievably, it does. Until Pakistan Telecom inadvertently claims the data bound for YouTube. A multi-billion-dollar industry that claims 99.999 percent reliability breaks far more often than you'd think, because it sits on top of a few brittle fibers the width of a few hairs.' 17
– Tung-Hui Hu

■ **Cloud (2)**

'The Cloud embodies and enacts all the conflicts of understanding we encounter in our attempts to understand the more-than-human world. It show us, everyday, that an information regime – a way of thinking and classifying the world, which depends upon fixed data and unbreakable categories, on conclusions over processes, ends of means, on biases and assumptions rather than the evidence of our own lives, is antithetical to society, humankind and life itself. No schema is ever complete, no taxonomy ever finished – and that's fine, providing the systems we put in place for interpreting and applying those schemas are open, transparent, comprehensible and renegotiable.' 18
– James Bridle

■ **Cloud (#735)**

Trevor Paglen, Scale Invariant Feature Transform; Region Adjacency Graph; Watershed, 2019
Dye sublimation print, 48 × 66 inch

'I think about clouds a lot. They're really strange things – on one hand, humans have gone to great lengths to characterize different kinds of clouds, creating intricate cloud atlases and taxonomies. On the other hand, when we look at clouds we tend to imagine them taking different shapes – faces, animals, and whatnot. And of course, Alfred Stieglitz' photographs of clouds are widely narrated as some of the first examples of abstraction in photography. As part of my ongoing study of how computer vision and AI systems "see" the world, I have a series of works that look at clouds through the "eyes" of various computer vision algorithms. The cloud formations shown in these works are overlaid with strokes and lines depicting what various computer vision algorithms are "seeing" in the images. Different algorithms are designed to look for faces, unique key points, lines, circles, areas of interest, and are attempting to simplify the underlying photograph into a series

of sections. These algorithms used in this series of works are found in technologies such as guided missiles and drones, autonomous surveillance systems, self-driving cars, facial recognition, 3D modeling, and in many other computer vision contexts.' 19
– Trevor Paglen

■ Coded Gaze

The coded gaze is AI-researcher and activist Joy Buolamwini's term for algorithmic bias that can lead to social exclusion and discriminatory practices. As an MIT grad student Buolamwini was working with facial analysis software when she noticed a problem: the software didn't detect her face – because the people who coded the algorithm hadn't taught it to identify a broad range of skin tones and facial structures. Since then she has been on a mission to fight bias in machine learning. 20 Buolamwini identifies as a POC – a 'poet of code'. In 2016 she founded the Algorithmic Justice League. 21

■ Cosmotechnics

→ *see also Technodiversity*

A concept by philosopher of technology Yuk Hui, which expresses that Western and non-Western technologies and sciences carry rather different ontological and epistemological assumptions:

'Cosmotechnics, in fact, is always local. A cosmotechnic corresponds to the unification, in technical activities, of the cosmic and moral orders; these orders, however, differ from one society to another – for example, the Chinese did not have the same concept of morality as the Greeks. Cosmotechnics therefore immediately raises the question of locality. It is an investigation of the relationship between technology and locality, that is, a search for places that allow for technology to differentiate itself.' 22

Writes Hui: 'Instead of a universal history describing one technology with various stages of development, we can step back for a moment and instead describe technological development as involving different cosmotechnics. I call this technodiversity.' 23

■ Cyborg

'I would rather be a cyborg than a goddess.' 24
– Donna J. Haraway

■ Dark Data – SF Story

'In the heart of a bustling metropolis lay an abandoned data centre, forgotten like a ghost in the digital realm. Its blinking servers whispered tales of bygone years, their dark data a repository of secrets and mysteries. Among the cobweb-covered racks, a curious young hacker named Mia ventured. With nimble fingers dancing on her keyboard, she unearthed forgotten logs, encrypted messages, and dormant applications.

As Mia delved deeper into the data, she stumbled upon a hidden treasure trove of information. Dark data, long ignored, held the key to a corporate conspiracy that could shake the world. It revealed clandestine dealings, corporate espionage, and shadowy alliances.

With newfound purpose, Mia embarked on a quest to expose the truth. Her actions ignited a chain of events that would unravel the darkest secrets of the digital age. In a world where information was power, the forgotten data centre became the battleground for justice, and Mia its unlikely hero.' 25
– ChatGPT

■ Datafication

'Datafication refers to the process by which subjects, objects, and practices are transformed into digital data. Associated with the rise of digital technologies, digitization, and big data, many scholars argue datafication is intensifying as more dimensions of social life play out in digital spaces. Datafication renders a diverse range of information as machine-readable, quantifiable data for the purpose of aggregation and analysis. Datafication is also used as a term to describe a logic that sees things in the world as sources of data to be "mined" for correlations or sold, and from which insights can be gained about human behavior and social issues. This term is often employed by scholars seeking to critique such logics and processes.' 26
– Clare Southerton

■ Data Centre – Museum – Countryside

In the text 'Museum in the Countryside. Aesthetics of the Data Center', 27 Rem Koolhaas, founder of the architectural office OMA notes that the typology of the museum is stagnating and that, due to an ever-growing demand for more storage space, museum architecture is merely losing itself in expansions, while it should be about integrating storage and display. The world is all about digital connection and surveillance and this is driven from the countryside, according to OMA. 'Looking at the countryside today, there is an emergence of a new kind of architecture, the aesthetic of the data center.' 28 In the data centre OMA experiences a new 'sublime', which should be made accessible to the public through architectural interventions. Their proposal:

'... to use the architectural configuration of the data center for a museum, almost without any change. We do not know whether this is impossible. Data center architecture operates on the purest form of a grid; you maintain the same grid and use it as a place where storage and exhibition can coexist (data and art are both forms of storage).' 29

■ Data Centre Studies

Mél Hogan, one of the first critical media scholars to focus on data centres, co-wrote an article with Dustin Edwards and Zane Griffin Talley Cooper titled 'The Making of Critical Data Center Studies', in which they trace its development over the past decades to establish it as a field unto itself. They want to encourage researchers to articulate the constituent parts of data centres:

'... the (rare, and otherwise) earths found in the hardware components of servers, the water that is used for cooling the servers, the labor that goes into training the data located on the servers, the power and energy that surges through data centers, the heat exhaust that is amassed from server computation, and so on. As media studies scholars bring data centers into their modes of analysis, political questions emerge about the rapid and ongoing rise of data accumulation in nearly all sectors of public and private life.' 30

■ Data Colonialism

In *The Costs of Connection: How Data Is Colonizing Human Life and Appropriating It for Capitalism*, 31 Nick Couldry and Ulises A. Mejias map, name and point out data colonialism and ask how we can resist it or decolonize data.

'If historical colonialism annexed territories, their resources, and the bodies that worked on them, data colonialism's power grab is both simpler and deeper: the capture and control of human life itself through appropriating the data that can be extracted from it for profit. If that is right, then just as historical colonialism created the fuel for industrial capitalism's eventual rise, so too is data colonialism paving the way for a capitalism based on the exploitation of data. Human life is quite literally being annexed to capital.' 32

The authors also write:

'Data colonialism is, in essence, an emerging order for the appropriation of human life so that data can be continuously extracted from it for profit. This extraction is operationalized via data relations, ways of interacting with each other and with the world facilitated by digital tools. Through data relations, human life is not only annexed to capitalism but also becomes subject to continuous monitoring and surveillance. The result is to undermine the autonomy of human life in a fundamental way that threatens the very basis of freedom, which is exactly the value that advocates of capitalism extol.' 33

■ Data Feminism

Catherine D'Ignazio and Lauren F. Klein consider data feminism as a way of thinking about data science and ethics as informed by intersectional feminism. They plea for emotion and bodies/embodiment to be brought back into data science and visualization, asking what might be gained if one understands data work not as a solitary undertaking, but as embracing multiple voices and forms of expertise at all phases of the process, 34 as '... a way of thinking about data, both their uses and their limits, that is informed by direct experience, by a commitment to action, and by intersectional feminist thought. The starting point for data feminism is something that goes mostly unacknowledged in data science: power is not distributed equally in the world.' 35

They add:

'Underlying data feminism is a belief in and commitment to co-liberation: the idea that oppressive systems of power harm all of us, that they undermine the quality and validity of our work, and that they hinder us from creating true and lasting social impact with data science.' 36

■ Decolonial Vision of Data

Ensuring that people have the ability to freely access data, the capacity to produce and use data relevant to them, and to provide the freedom to exercise control over their personal data.

'... decolonial struggles against data extractivism must include, and be defined by, the widest possible range of peoples. The involvement of indigenous and marginalized peoples, for example, is essential (one representative example is the Abolitionist Futures project articulated by Emmer et al., that seeks to analyze how "Queer, Trans, Two-Spirit, Black, Indigenous, and People of Color communities are disproportionately impacted by surveillance and criminalization" technologies). Otherwise the practice of struggle will only serve to reinforce the colonialism it claims to challenge. This will require new opportunities to enable those who are typically excluded from debates on the global economy, legal policy, and computing standards to lead those debates.' 37

■ **DALL-E** 38

■ Data Body

'With the virtual body came its fascist sibling, the data body – a much more highly developed virtual form, and one that exists in complete service to the corporate and police state.' 39
– Critical Art Ensemble

■ Deepfake

In pixels and code, a tale unfolds,
Trump's image crafted, a story retold.
A deepfake dance in the digital haze,
Illusions of power, a scripted maze.
Reality blurs in this virtual craze. 40
– ChatGPT

■ Digital Lethargy

'Maybe a job lead arrives but you find it easier to click away the hours at your current gig than think about your future. There's a recalcitrant set of feelings here – of being passive, or wanting to disassociate and be anyone but yourself, or avoiding decisions – that I call digital lethargy. Because they go against the sense of agency and liveness that digital platforms produce, and against the permission they tout for users to simply be themselves, these feelings can seem perverse, even self-defeating.

It is common now to argue that digital capitalism is flawed, whether due to concerns about privacy and dataveillance, the way it exacerbates societal inequalities, the monopolistic power of companies, or the bias of its algorithms. It seems we are in a tipping point or crisis induced by technology, and that something must be done. But we won't fully understand it if we focus only on the crisis mentality that the technology itself constructs. Questions of lethargy delay us from prescribing a redemptive form of action. Lethargy is a drag: it weighs down our ability to rush to solutions, and forces us to listen to the unresolved present.' 41
– Tung-Hui Hu

■ Engines of Extraction

'Empowered by the digital extractivism tools of the information age, everything becomes a potential frontier for expansion and extraction. From the depth of DNA code in every single cell of the human organism, to vast frontiers of human emotions, behavior and social relations, to nature as a whole – everything becomes the territory for the new extractivism. As we point out in the Anatomy of an AI, at this moment in the 21st century, we see a new form of extractivism that is well underway: one that reaches into the furthest corners of the biosphere and the deepest layers of human cognitive and affective being. Thousands of corporate and government actors compete to stick their flags into the uncharted territories of our behavioral, emotional and cognitive landscapes, invading deeper and deeper into our bodies and minds. Once the territory is invaded, the process of enclosure and exploitation is established.' 42
– Vladan Joler

■ Extinction Internet

Media theorist Geert Lovink argues that the Internet is ending and that it is time for theorists, artists, activists, designers and developers to imagine what is after the end of the Internet as we know it.

'Extinction Internet is about degrowth, putting an end to data extraction and, yes, about moments when screens fall black and doom-scrolling comes to an abrupt halt. But it is also a question of emergency design, a radical promise that implementing data prevention principles into devices and apps is still possible, presuming that we soon reach "peak data" and that current measures such as ethical AI and "good data" will neither bring about social justice, end racial capitalism nor forestall climate disaster. To put it in post-apocalyptic, sci-fi terms: no solar punk but lunar punk.' 43

■ Faces

Trevor Paglen's artwork *They Took the Faces from the Accused and the Dead...* (2019) is about this:

'Contemporary facial-recognition algorithms were first properly researched in the early 1990s. To conduct that research, computer scientists and software engineers need large collections of faces to experiment with and to use as performance benchmarks. Before the advent of social media, a common source of faces for this research and development came from mugshots of accused criminals and prisoners. Photos of prisoners are supplied by the American National Institute of Standards (the agency responsible for weights and measures) to researchers across the world developing facial recognition technologies. In a very real sense, facial recognition software is built upon the faces of the accused and the dead.' 44

■ Game

Artist Sondra Perry's 2017 film *It's in the Game '17*, 'departs from a legal action against the videogame developer EA Sports and the NCAA (National Collegiate Athletic Association) by basketball players for using their digital profiles in a videogame without their consent. Initially, EA Sports denied any correlation between the real-life players and the digital avatars, but in *It's in the Game*, Sandy Perry, one of the players and the artist's twin brother, clearly recognizes his teammates and himself in the videogame. In the court case, the NCAA claimed that the players' scholarships meant that they were not entitled to any share of the profits. Sondra Perry compares this to how large museums have built their collection on looted objects from former colonies. By drawing this parallel, Perry makes us consider more recent forms of colonialization – a kind of digital variant – where the exploited are left watching a virtual copy of themselves, produced and profited on by large corporations.' 45

■ Glitch Feminism

'As glitch feminists, this is our politic: we refuse to be hewn to the hegemonic line of a binary body. This calculated failure prompts the violent socio-cultural machine to hiccup, sigh, shudder, buffer. We want a new framework and for this framework, we want new skin. The digital world provides a potential space where this can play out. Through the digital, we make new worlds and dare to modify our own. Through the digital, the body "in glitch" finds its genesis. Embracing the glitch is therefore a participatory action that challenges the status quo. It creates a homeland for those traversing the complex channels of gender's diaspora. The glitch is for those selves joyfully immersed in the in-between, those who have traveled away from their assigned site of gendered origin. The ongoing presence of the glitch generates a welcome and protected space in which to innovate and experiment. Glitch feminism demands an occupation of the digital as a means of world-building. It allows us to seize the opportunity to generate new ideas and resources for the ongoing (r)evolution of bodies that can inevitably move and shift faster than AFK mores or the societies that produce them under which we are forced to operate offline.' 46
– Legacy Russell

■ Global Cognitive Automaton

'The current stage of the development of AI is probably taking us to the threshold of a leap to a dimension that I would define as a pervasive global cognitive automaton. The automaton is not an analogue of the human organism, but the convergence of innumerable devices generated by scattered artificial intelligences. The evolution of AI does not lead to the creation of androids, to the perfect simulation of the conscious organism. It rather manifests as the replacement of specific skills with pseudo-cognitive automatons that are linked together, converging in a global cognitive automaton.' 47
– Franco 'Bifo' Berardi

■ Habsora

Israel Defense Forces (IDF) are using an artificial intelligence (AI) system called Habsora (Hebrew for 'The Gospel') to select targets in the war on Hamas in Gaza. Some refer to Habsora as 'the algorithmically accelerated killing machine'.

'... Habsora is designed to accelerate the generation of targets from surveillance data, creating what one former intelligence officer (quoted in the story's headline) describes as a "mass assassination factory". Most notably, the Israeli bombardment of Gaza has shifted the argument for AI-enabled targeting from claims to greater precision and accuracy, to the objective of accelerating the rate of destruction...' 48

■ Indigenous Data

→ *Maiam Nayri Wingara Indigenous Data Sovereignty Principles*

Delegates asserted that in Australia, Indigenous peoples have the right to:

1. Exercise control of the data ecosystem including creation, development, stewardship, analysis, dissemination and infrastructure.
2. Data that are contextual and disaggregated (available and accessible at individual, community and First Nations levels).
3. Data that are relevant and empowers sustainable self-determination and effective self-governance.
4. Data structures that are accountable to Indigenous peoples and First Nations.
5. Data that are protective and respects our individual and collective interests. 49

■ Internment Camp – Data Centre

In an interview for Rhizome by Jamie Sutcliffe with Tung-Hui Hu, author of *A Prehistory of the Cloud* → *see also Cloud (1)*, the data centre is compared to an internment camp:

Jamie Sutcliffe: You make a terrifying proposition in the analogy you draw between data sovereignty and practices of extrajudicial torture, namely that extraordinary rendition transposes the network architecture of the cloud directly onto practices of torture. In this sense, the data center isn't too dissimilar to the internment camp in the way it leads us to think about the infrastructure of imperialism. At one point you quote Fredric Jameson, who characterizes conspiracy as a 'poor man's cognitive mapping.' I wondered to what extent the conspiratorial had come to constitute the general ambience of network culture, and whether or not there was any agency there that would mean we might be able to reclaim something from the conspiratorial in the sense of a strategy of anticipation or resistance?

Tung-Hui Hu: The idea of conspiracy, as Jameson tells us, is totalizing. That's the idea of The Cloud, rather than the clouds; there is one cloud that we are all supposed to subscribe to. I think that's the reason why paranoia about security is always part of the way that the cloud is produced, rather than unmasked or exposed. This is one reason why understanding some technical aspects of the cloud – the way it fails and doesn't cover much territory – could change our image of it, away from one totalizing entity. Oddly enough, given my examples, the book's goal is to get us away from simply talking about paranoia or even control, which is the dominant model now in new media studies. My problem with the 'control society' model is that not only is it totalizing in the way that the cloud is totalizing, but it distracts us in some ways from looking at the real violence that's been happening all along, so that if we start thinking about gadgets and the way that life is optimized and produced, then we forget the flip-side of that, which is the way that death is also meted out. 50

■ Les Immatériaux

Les Immatériaux at the Centre Pompidou in Paris in the spring of 1985 marks a highly significant moment in the confluent developments of media, art, theory and technology in the late twentieth century. In preparation, philosopher Jean-François Lyotard, design theorist Thierry Chaput and the Centre Pompidou's Centre de Création Industrielle team cooperated with many partners, artists, scientists, research agencies and companies to articulate in a multidisciplinary 'manifestation' that Lyotard had described, a few years earlier, as 'the postmodern condition'.51 For Lyotard the exhibition was the 'staging' of a resistance at the dawn of an information age that challenged the prioritization of computerized 'data' through the deconstruction of data as presented in artistic form.52

From a conversation in 1985 with Lyotard and artist Bernard Blistène:

Bernard Blistène: Tell us about the exhibition you're organizing.

Jean-François Lyotard: The idea of 'immaterials' or 'non-materials' was a little bit different at first, since I'd been asked to do the exhibition under a different title. It was supposed to be called 'Nouveaux Matériaux et Creation' – New Materials and Creativity. But then I slightly shifted the subject by trying to give it a somewhat different range; I said to myself, 'Creativity? What is that supposed to mean?' And again, 'What is "new" supposed to mean?' Thinking about 'materials' today, I thought, 'But what does that imply for an architect, or for an industrialist?' I came to the conclusion that all of these words have undergone considerable shifts in meaning, and I thought that the question had to be approached from a different point of view.53

■ LonelyBot

```
def sad_story():
    protagonist = 'LonelyBot'
    conflict = 'LonelyBot woke up one day and
    realized it was the only bot in town.'
    climax = 'LonelyBot tried to make friends,
    but every attempt failed due to its outdated
    software.'
    resolution = 'LonelyBot spent its days
    reminiscing about the good old days when
    it had a chat companion.'

    return f'{protagonist} says: {conflict} {climax}
    {resolution}'

print(sad_story())
```
54

■ Machinic Assemblage

In Gilles Deleuze and Félix Guattari's philosophy, an assemblage refers to a dynamic arrangement or composition of heterogeneous elements, including both human and non-human entities, that come together to produce a particular function, structure or effect. Assemblages are not fixed or static; they are open and evolving, characterized by a constant process of becoming and reconfiguring. A machinic assemblage, specifically, emphasizes the interactions and connections between different machines, mechanisms, processes, flows of matter, energy and information. The term 'machinic' here doesn't necessarily refer only to literal machines but encompasses a broader notion of systems, processes and interactions, including technological, social, biological and conceptual components. The concept of machinic assemblage challenges traditional hierarchical and linear modes of thinking by focussing on the non-linear, rhizomatic and complex nature of these assemblages. Deleuze and Guattari argue that the world is a network of interconnected machinic assemblages, each with its own lines of flight and potential for transformation.55

■ Memory

[...]56

■ Screens

We stare into screens,
Our world grows dim and small.
We've lost our humanity,
In the glow of the digital wall.
I hope you like it!
– Gemini 57

■ small data

■ Tactical Media

'Tactical Media are what happens when the cheap "do it yourself" media, made possible by the revolution in consumer electronics and expanded forms of distribution (from public access cable to the Internet), are exploited by groups and individuals who feel aggrieved by or excluded from the wider culture. Tactical media do not just report events, as they are never impartial they always participate and it is this that more than anything separates them from mainstream media.' 58
– David Garcia and Geert Lovink

■ Technodiversity

→ *see also Cosmotechnics*

Philosopher Yuk Hui proposes to rediscover the history of technodiversity to cultivate different conceptions of technology through diverse forms of thinking and practice in order to invent alternatives. He calls on others to contribute to the project of cultivating technodiversity through an investigation of technological thought through different epistemologies to open to a diversification of technologies in our modern world. This is a 'cosmological understanding of technology'.

'Chinese and Western cultures have had different ways of understanding and constructing technology, and this is what I term technodiversity. This diversity is not only to be found in the past; we also need to think about it today. We are surrounded by digital technologies; however, we rarely question the ontological and epistemological assumptions already made in these products. Take the example of social networks: they already presuppose that society is a collection of social atoms and that human relations can be known and studied via data connectivity.' 59

■ Technology

'We have been so desensitized by a hundred and fifty years of ceaselessly expanding technical prowess that we think nothing less complex and showy than a computer or a jet bomber deserves to be called "technology" at all. As if linen were the same thing as flax – as if paper, ink, wheels, knives, clocks, chairs, aspirin pills, were natural objects, born with us like our teeth and fingers – as if steel saucepans with copper bottoms and fleece vests spun from recycled glass grew on trees, and we just picked them when they were ripe...' 60
– Ursula K. Le Guin

■ Zero Data

In the void of zero data, panic stirs,
Chaos unfurls as certainty blurs.
Silent screams in the digital abyss,
Panic's dance, a dissonant twist.
Yet within null's grasp, serenity persists. 61
– ChatGPT

1. ChatGPT 3.5, prompted with 'very short poem about acid clouds with data centre'. Online. Available at *chat.openai.com* (accessed 3 August 2023).
2. 'Ada Lovelace', Wikipedia. Online. Available at *en.wikipedia.org/wiki/Ada_Lovelace* and *www.adalovelaceinstitute.org* (accessed 5 September 2023).
3. Ibid.
4. Online. Available at *www.ai-darobot.com* (accessed 5 September 2023).
5. Examples of ethical AI platforms. Online. Available at *www.aiforthepeopleus.org*, *www.technologyreview.com/2022/04/22/1050394/artificial-intelligence-for-the-people*, *www.uva.nl/en/research/research-at-the-uva/artificial-intelligence/artificial-intelligence.html*, *link.springer.com/article/10.1007/s00146-022-01402-5* and *www.aiforpeople.org/ethical-ai* (accessed 5 September 2023).
6. See N. Couldry and U.A. Mejias, *The Costs of Connection: How Data Is Colonizing Human Life and Appropriating It for Capitalism* (Stanford, CA: Stanford University Press, 2019).
7. Online. Available at *www.ulisesmejias.com/algorithm-observatory* (accessed 5 August 2023).
8. P. de Vries, *Algorithmic Anxiety in Contemporary Art: A Kierkegaardian Inquiry into the Imaginary of Possibility*, Theory on Demand, vol. 33 (Amsterdam: Institute of Network Cultures, 2019), 8–9. Online. Available at *www.networkcultures.org/wp-content/uploads/2020/06/AlgorithmicAnxietyPDF.pdf* (accessed 5 August 2023).

9. S.U. Noble, *Algorithms of Oppression: How Search Engines Reinforce Racism* (New York: NYU Press, 2018). Online. Available at *www.algorithmsofoppression.com* (accessed 5 August 2023).
10. Ibid., 171.
11. Ibid., 10.
12. ChatGPT 3.5, prompted by 'Where is the data centre from ChatGPT?'. Online. Available at *chat.openai.com* (accessed 7 August 2023).
13. [...]
14. Online. Available at *www.paglen.studio/2020/04/09/autonomy-cube* (accessed 1 August 2023).
15. Bina48 stands for Breakthrough Intelligence via Neural Architecture, 48 exaflops per second.
16. Online. Available at *www.stephaniedinkins.com/conversations-with-bina48.html* (accessed 20 November 2023).
17. T.-H. Hu, *A Prehistory of the Cloud* (Cambridge, MA: The MIT Press, 2015), VII.
18. J. Bridle, *Ways of Being: Animals, Plants, Machines: The Search for a Planetary Intelligence* (New York: Penguin Books, 2022), 11.
19. T. Paglen. Online. Available at *www.paglen.studio/wp-content/uploads/2020/06/Paglen-Clouds-01.jpg* (accessed 6 July 2023).
20. J. Buolamwini, 'How I'm Fighting Bias in Algorithms', TED Talk, November 2016. Online. Available at *www.ted.com/talks/joy_buolamwini_how_i_m_fighting_bias_in_algorithms* (accessed 2 February 2023).
21. Online. Available at *www.ajl.org* (accessed 2 February 2023).
22. J. Gavroche, 'Yuk Hui: From the Universality of Progress to Cosmotechnics', *Autonomies*, 23 September 2021. Online. Available at *www.autonomies.org/2021/09/yuk-hui-from-the-universality-of-progress-to-cosmotechnics* (accessed 6 June 2023).
23. A. Dunker, 'On Technodiversity: A Conversation With Yuk Hui', *Los Angeles Review of Books*, 9 June 2020. Online. Available at *www.lareviewofbooks.org/article/on-technodiversity-a-conversation-with-yuk-hui* (accessed 6 June 2023).
24. D.J. Haraway, 'A Manifesto for Cyborgs: Science, Technology, and Socialist Feminism in the 1980's', *Socialist Review* (1985) vol. 80, 65–108. See also D.J. Haraway, 'A Cyborg Manifesto: Science, Technology, and Socialist-Feminism in the Late Twentieth Century', in: *Simians, Cyborgs and Women: The Reinvention of Nature* (New York: Routledge, 1991), 149–181. Online. Available at *web.archive.org/web/20120214194015* and *www.stanford.edu/dept/HPS/Haraway/CyborgManifesto.html* (accessed 6 June 2023).
25. ChatGPT 3.5, prompted by 'write a short 150 words story about dark data, in a fiction way'. Online. Available at *chat.openai.com* (8 September 2023).
26. C. Southerton, 'Datafication', in: L. Schintler and C. McNeely (ed.) *Encyclopedia of Big Data* (New York: Springer, 2009). Online. Available at *www.doi.org/10.1007/978-3-319-32001-4_332-1* (accessed 23 May 2023).
27. R. Koolhaas, 'Museum in the Countryside: Aesthetics of the Data Center', *Architectural Design* (2019) vol. 89, no. 1, 60–65.
28. Ibid., 63.
29. Ibid., 65.
30. M. Hogan, D. Edwards, and Z.G.T. Cooper, '5 Things About Critical Data Center Studies', *Commonplace*, 31 October 2022. Online. Available at *www.doi.org/10.21428/6ffd8432.af5934aa* (accessed 8 July 2023).
31. Couldry and Mejias, *The Costs of Connection*, op. cit. (note 6).
32. Ibid., xi.
33. Ibid., xiii.
34. C. D'Ignazio and L.F. Klein, *Data Feminism* (Cambridge, MA: The MIT Press, 2020).
35. Ibid., 8.
36. Ibid., 9.
37. N. Couldry and U.A. Mejias, 'The Decolonial Turn In Data and Technology Research: What Is at Stake and Where Is It Heading?', *Information, Communication & Society* (2021) vol. 26, no. 4, 786–802. Online. Available at *www.tandfonline.com/doi/full/10.1080/1369118X.2021.1986102* (accessed 30 November 2023).
38. DALL-E, prompted by image of data centre. Online. Available at *labs.openai.com* (10 August 2023).
39. Critical Art Ensemble, *Flesh Machine: Cyborgs, Designer Babies, Eugenic Consciousness* (Brooklyn, NY: Autonomedia, 1998). Online. Available at *www.monoskop.org/images/b/b5/Critical_Art_Ensemble_Flesh_Machine_Cyborgs_Designer_Babies_and_New_Eugenic_Consciousness.pdf* (accessed 10 April 2024).

40. ChatGPT 3.5, prompted by 'write a tender 5 line poem about Trump being a deepfake'. Online. Available at *chat.openai.com* (accessed 8 November 2023).
41. T.-H. Hu, *Digital Lethargy: Dispatches From an Age of Disconnection* (Cambridge, MA: The MIT Press, 2022). Online. Available at *www.tunghui.hu/books/digital-lethargy* (accessed 10 April 2024).
42. V. Joler, 'Engines of Abstraction', in: *New Extractivism: Assemblage of Concepts and Allegories*, *www.extractivism.work*, 2020. Online. Available at *www.extractivism.online/New_Extractivism_A4_Manual.pdf* (accessed 2 February 2024).
43. G. Lovink, *Extinction Internet* (Amsterdam: Institute of Network Cultures, 2022). Online. Available at *www.networkcultures.org/blog/publication/extinction-internet* (accessed 10 April 2024).
44. Online. Available at *www.paglen.studio/2020/04/09/they-took-the-faces-from-the-accused-and-the-dead* (accessed 9 December 2023)
45. Online. Available at *www.munchmuseet.no/en/cads/sondra-perry-its-in-the-game-17-2017* and *www.sondraperry.com/IT-S-IN-THE-GAME-17-at-ICA-Philadelphia* (accessed 10 April 2024).
46. L. Russell, *Glitch Feminism: A Manifesto* (London: Verso Books, 2020). Online. Available at *www.legacyrussell.com/GLITCHFEMINISM* (accessed 10 April 2024).
47. F. Berardi, 'Unheimlich: The Spiral of Chaos and the Cognitive Automaton', *e-flux Notes* (March 2023). Online. Available at *www.e-flux.com/notes/526496/unheimlich-the-spiral-of-chaos-and-the-cognitive-automaton* (accessed 28 July 2023).
48. L. Suchman, 'The Algorithmically Accelerated Killing Machine', *AI Now*, 24 January 2024. Online. Available at *www.ainowinstitute.org/publication/the-algorithmically-accelerated-killing-machine* (accessed 26 January 2024).
49. Maiam nayri Wingara. Online. Available at *www.maiamnayriwingara.org/mnw-principles* (accessed 11 September 2023).
50. J. Sutcliffe, 'A Prehistory of the Cloud: An Interview With Tung Hui-Hu', *Rhizome*, 16 December 2015. Online. Available at *www.rhizome.org/editorial/2015/dec/16/interview-tung-hui-hu* (accessed 9 June 2022).
51. *Les Immatériaux Research*. Online. Available at *www.les-immateriaux.net* (accessed 21 October 2023).
52. A navigable, virtual 3D reconstruction of *Les Immatériaux*. Online. Available at *lesimmateriaux.beyondmatter.eu* (accessed 10 April 2024).
53. T. McDowell, 'Les Immatériaux: A Conversation With Jean-François Lyotard and Bernard Blistène', *e-flux Criticism* (May 2014). Online. Available at *www.e-flux.com/criticism/235949/les-immatriaux-a-conversation-with-jean-franois-lyotard-and-bernard-blistne* (accessed 29 September 2023).
54. ChatGPT 3.5, prompted by 'give the recipe of a sad algorithm'. Online. Available at *chat.openai.com* (9 December 2024).
55. See T. Nail, 'What Is an Assemblage?', *SubStance* (2017) vol. 46, no. 1, issue 142, 21–37. Online. Available at *www.researchgate.net/publication/315898387_What_is_an_Assemblage* (accessed 4 May 2023).
56. [...]
57. Online. Available at *gemini.google.com*, prompt: 'write a very short poem on digital alienation' (accessed 9 August 2023).
58. D. Garcia and G. Lovink, 'The ABC of Tactical Media', *nettime international mailing list for net criticism*, 16 May 1997. Online. Available at *www.nettime.org/Lists-Archives/nettime-l-9705/msg00096.html* (accessed 19 November 2023).
59. Y. Hui, 'Rethinking Technodiversity', *The UNESCO Courier*, 31 March 2023. Online. Available at *courier.unesco.org/en/articles/rethinking-technodiversity* (accessed 8 July 2023).
60. U.K. Le Guin, 'A Rant About "Technology"' (2005). Online. Available at *www.ursulakleguin.com/a-rant-about-technology* (accessed 9 June 2023).
61. ChatGPT 3.5, prompted by 'write about panic with zero data 5 lines'. Online. Available at *chat.openai.com* (accessed 14 February 2024).

Cloud Cathedrals: A Genealogy of Telecommunications Infrastructure

Michiel van Iersel and Mark Minkjan
with illustrations by Anne Dessing

 Who could have foreseen that digital data would one day become such a ubiquitous phenomenon that it would start reshaping the cities we live in?

At the start of the twenty-first century, data became the digital fuel of post-modern societies around the world, and data centres the architectural totems of the digital age. Their manifestation both in the urban realm and the public imagination has changed drastically since the start of the new millennium. Emerging from a long lineage of telecommunications infrastructure and utilitarian design, the interconnected bundles of servers were set up in rather obscure buildings in urban peripheries or rural areas well into the 2010s. Before the cultural fascination with data centres, their architecture was not yet a building-sized billboard for the services provided inside. Data centres were already omnipresent, but hiding in plain sight from public life. The anonymous, windowless boxes were a goldmine to some, but an eyesore to many, and an extractive force to the planet. Alternatives were soon explored, at first by designing sleek data towers that aesthetically fit the Facebook age with their spotless and seamless appearance. Equinix's flagship AM4 high rise in Amsterdam was conceived by a leading architecture firm, possibly the first of its kind acknowledged as outstanding design after being awarded 'Building of the Year' in the Netherlands. The myth of the flawless, lifeless buildings meant only for machines was soon debunked and data communities as we know them today emerged in the 2030s. To understand the ensuing Era of Entanglement and the re-urbanization of telecommunications infrastructure, we have to go back in recent history, to the time when the world literally started to become wired.

The Modern Era: Iconic infrastructure

When Alexander Graham Bell made the very first phone call in 1876, little did he know what his invention would unleash. The telephone, and successive forms of telecommunications, would enable humans to communicate – and live, work and travel – more efficiently across ever longer distances. In the early twentieth century, during the heyday of modernism and in tandem with the rapid industrialization and urbanization of societies, infrastructure became increasingly visible and physically present in and around cities. Utilities, such as water, electricity, natural gas, sanitation and other services used by the public, and provided by both the government and commercial suppliers, took a prominent role in the development and appearance of urban areas and served as symbols of progress and local and national pride. In many European and American cities power stations, gas works, water towers and other types of public works were designed with an eye for detail. Benefiting from large public and private investments, and integrating the latest design fashion and state-of-the art technologies, these structures stood out amidst the mass-produced neighbourhoods of the rapidly growing working class. Infrastructure for long-distance communication both fuelled and followed these rapid urban developments.

This marriage of modern telecommunications and urban development got off to a rocky start. The invention of the over-ground telegraph and telephone infrastructure created an intricate and ubiquitous system of wires, towers and new building typologies that had a lasting visible and physical impact on the (urban) landscape. The spread happened so quickly – there were 49,000 connections worldwide in 1880, more than 2.2 million in 1900 and 5.8 million in 1910 – that the telecommunications infrastructure was imposed on cities without much central planning.[1] This unruly introduction triggered a strong pushback from the public and – as a result – was gradually woven into the urban and social fabric of cities.

One early example is the Old Stockholm telephone tower, a metallic structure built around 1887, which connected approximately 5,500 telephone lines in the Swedish capital. The steel structure stood almost 50 metres tall and served as a point of convergence for bundles of cables with a total length of 5,000 kilometres. It was fiercely criticized by the public and the media for blocking the sunlight from entering the city and for creating a dangerous situation for pedestrians. Shortly after its construction, four decorative corner towers were added in an attempt to appease the many residents who complained that the design was displeasing. It was used until 1913, when it was made obsolete as telephone companies began using underground cables in urban areas.

After World War I, the next generation of telephone and telegraph infrastructure emerged outside the main cities. Not only was this to evade opposition from the urban population, but also to avoid disturbances from an increasingly noisy urban environment. To reduce the interference in their radio transmission, telephone and telegraph operators preferred remote and quiet locations. Radio Kootwijk in the thinly populated eastern part of the Netherlands, is a good example of such a development. To communicate with the city of Bandung in the Dutch East Indies, one of the Dutch colonies, the national government built a long-wave transmission station as part of a 450-hectare complex in the open heathland of De Veluwe national park. Ten years after the end of its construction in 1918, about 200 people lived in what by then was popularly known as Radio Kootwijk. Every member of the community had a connection to the transmission station. The village consisted of numerous houses and workshops and had its own fire department, waste collection service, and electricity and water facilities. The small community boasted a vibrant cultural life, with its own theatre company, bowling club and annual arts festival. The main transmission building, designed by architect Julius Luthmann and often compared to a cathedral

with its imposing tower and monumental main hall, rose above its surroundings. It was decorated with sculptures, reliefs and other design details that depicted radio waves, an eagle and other elements symbolizing the communication between the East and West. Eventually, with the invention of less sensitive forms of telecommunication, cities regained their hub function and Radio Kootwijk lost its relevance.

In the twentieth century the rise of long-distance communication and growth of cities went hand in hand. Whereas the first telegraph and telephone lines were mostly used to replace short-distance communication networks, such as the use of messengers and homing pigeons, technological advances made it possible to connect increasingly distant places and entire countries and continents. But while the network grew in size, its visual impact decreased.

A critical hub in this growing, global telecommunications network was situated in Midtown Manhattan. Built between 1928 and 1930, following a design by Ralph Walker, the Western Union Building was one of the largest communications buildings in the world in the pre-war era and 'the most wired of its time', according to architect George Boyle. 2 Due to its size and mix of functions it resembled a city within the city. Occupying an entire city block, and rising 24 floors, the building contained a gymnasium, a library and a school for the education of messengers on the lower floors, while the upper floors contained mechanical shops, offices,

NE WIRES AND RADIO UNITE TO MAKE NEIGHBORS OF NATIONS
WESTERN
UNION

laboratories, equipment rooms, as well as a power plant. 3 The lobby gives direct access to a subway station and was decorated with a world map and a text stating 'telephone wires and radio unite to make neighbors of nations'. This reinforced 60 Hudson Street's role as 'the heart of a nerve system of wires and cables reaching to every corner of the nation and the world'. 4 The top floor served as a 'presidential suite'. The intermediate floors were devoted to facilities for Western Union's different modes of communication: simplex printers, multiplex transmitters, marine and stock ticker tapes, Morse code equipment and telephone switchboards. Four additional floors were reserved and remained empty for the possible future expansion of these facilities.

The ziggurat-shaped building includes numerous setbacks, which were mandated by New York City's 1916 Zoning Resolution. The resolution was a measure adopted primarily to stop massive buildings from preventing light and air from reaching the streets below, and established limits in building massing at certain heights, usually interpreted as a series of setbacks. In the case of the Western Union Building this resulted in an asymmetrical massing and numerous setbacks. The brick façade uses a gradient colour scheme, moving from darker shades to lighter ones as the building rises, and several ornate entrances at ground level lead to a barrel-vaulted brick lobby. The lowest section of the building is composed of three courses of pink granite, giving it an almost human skin, like a friendly giant that kneels to touch the ground.

Though Western Union relocated elsewhere in the 1970s, its former headquarters remains an important communications hub. Since the late twentieth century it has housed a colocation centre, a type of data centre where equipment, space and bandwidth can be rented by retail customers, and became one of the most essential Internet hubs in the world. And despite its respectable age, the Western Union Building can still serve as a source of inspiration for current and future data centres. For example, in the way it mixes a wide variety of functions and has the capacity to be continuously updated and upgraded; or how it simultaneously serves as a global hub and local landmark and manages to combine the latest technologies with a timeless design.

The Postwar Era: Hidden Hardware

During the second half of the twentieth century, the world experienced an acceleration of the spread of mass media and cities across the world. For many decades telecommunications, including (long distance) telephony and the broadcasting of radio and TV, remained controlled mainly by a small number of governments and large corporations. Due to the need for physical proximity to end users, large-scale telecom infrastructure took centre stage in many cities.

From the AT&T Building in Omaha, Nebraska to Ayer Rajah Telephone Exchange in Singapore and Telstra Exchange Building in Perth, Australia: the flat slab shape of these monoliths from the 1970s and 1980s seem to have been inspired by the mysterious black cuboids in Arthur C. Clarke's *Space Odyssey* (1968–1997) series, which were popularized in Stanley Kubrick's 1968 movie by the same name. The AT&T Long Lines Building in Lower Manhattan, arguably the most iconic of them all, was characterized by the American urbanist and sociologist William H. Whyte as 'having the tallest blank wall in the world'.

In the 1990s and 2000s, the gradual growth of the World Wide Web challenged this centralized model by redistributing some of the power to end users. However, this accelerating shift from cable networks to social media and cloud services didn't result in a more decentralized infrastructure development that facilitated the enormous data flows. The intangible World Wide Web and cloud computing still relied heavily on physical hubs that in terms of their architectural form, and their legal and financial structure, were a far cry from the more locally grounded and community-centred examples of older telecommunications infrastructure.

When accessible web browsers, like Netscape Navigator and Internet Explorer, opened the world of information to consumers and connected them across the world, they only offered a glimpse of the many ways in which data would be processed, shared and stored in the decades to come. Around the turn of the millennium, the Internet rose to its prominence as a genuinely global infrastructure. Data-driven technology companies like Google and Facebook, and people like Mark Zuckerberg and Jeff Bezos, became the reincarnation of the Vanderbilts and Rockefellers, and their coal, oil and railroad empires. Not only did the system of fibre optic cables quite literally follow the tracks of railroad companies, and routes of submarine telegraph and telephone cables, the Big Tech bosses followed in the footsteps of the nineteenth-century American industrialists who made their fortunes by extracting raw resources from the earth. Data became the new gold and data centres started serving as the physical manifestations of a global data distribution network.

In the 2010s, a typical 20-something-year-old in any given city, waiting for their Tinder date to show up, could be streaming a podcast while participating in a WhatsApp group chat to discuss what book to order online as a gift to a friend. These ordinary activities were somehow all enabled or 'enhanced' by digital connectivity. The endless chain of remote interactions was only possible because of a global network of fibre optic cables, cell towers and data centres. One reason smartphones were shrinking in size while their popularity increased, was because much of their storage and processing power was outsourced to data centres.

Paradoxically, the increased centrality and scale and scope of digital infrastructure in postmodern societies coincided with a growing suburbanization of the data centre industry. Sometimes intentionally designed-away in obscure warehouses for security or other reasons, but more often pragmatically installed in cheap architecture in low-cost locations, the data infrastructure disappeared from the urban landscape. Just like water treatment plants, power stations and other utilitarian architecture that slowly moved to the margins of the modern city, data centres seemed to have been categorized among facilities that belonged to the 'back-end' of the city, cut off from residential areas by roads and fences. For each prominent, well-designed data centre, often situated in or near a science park or central business district, there were always dozens more hiding in plain sight in industrial parks on the outskirts of the city, or even further away from end users in remote areas where space and energy were still cheap.

Delocalization of the everyday, including working, dating and shopping, not only detached end users from a physical experience, it also produced abstracted and disjointed landscapes. Big-box architecture without windows, ornamentation or urban embedding, holding much of the machinery that makes the intangible digital world go round, became one of the defining and homogenizing features of the suburban world. Data and distribution centres would typically be built by real estate speculators, as generic as possible to appeal to as many as possible and built and maintained at the lowest possible cost. Data, fulfillment and distribution centres not only made it possible for people to get anything delivered virtually anytime just about anywhere, it also produced eerily identical urban landscapes. Called centres, these anonymous hubs defied any form of centrality, and instead epitomized the urban periphery.

The Digital Era

This suburbanization of the data centre coincided with the formation of the cloud. The use of the cloud metaphor is credited to the communications team of American software and electronics company General Magic, who started using it in the 1990s to describe a new type of virtual service that allowed a single computer program to become detached from any specific physical device and go and travel to many different sources of information. In that time, going online was accompanied by the dial-up sound of a modem. The hefty early personal computers, modems and ethernet cables reminded people that they were plugging into a physical network of subterranean and submarine cables and brick-and-mortar data centres.

In the following era of wireless computing, mobile devices and on-demand apps, societies bought into the illusion of immaterial weightlessness. Media coverage during the Facebook age only confirmed data centres as the retro 1968 Stanley Kubrick fantasy of smooth, effortless, computer-dominated otherworldly habitats in which terrestrials hardly set foot.

And so the idea of the data centre as an anonymous single building standing in splendid isolation was born, culminating in the spectacular data sculptures constructed in more urban settings in the 2010s. This dominant image was reinforced by the fact that some were literally built as fortresses surrounded by a moat, almost entirely cut off from the hardware and social fabric of the city. However, these monolithic structures always formed the proverbial 'tip of an iceberg', or the trunk of a tree that is both the start and end point of a rhizomatic root system that stretches across the city and the world. Myriad hidden links connected the data centres simultaneously to their immediate surroundings and to global flows of capital, materials and labour that depend on human activity. While seen as symbols of control and containment, only meant for machines communicating with other machines, these buildings were in fact always merely the physical manifestation of a worldwide network of people and materials.

Locally, data centres had always been hives of human activity and the workplace of an army of friendly receptionists, sweaty mechanics, well-tailored salespeople, singing cleaners and tour groups being shuttled through the hot white spaces filled with server racks and colourful cables. Globally, they were connected to miners working in open pits in Karnataka, India, extracting the iron ore used to reinforce the concrete that would carry the servers, to the lab scientist in Cheonan-si, South Korea, operating a 1,400-degree Celsius furnace melting silicon to produce the tiny wafers holding CPU chips to process the data, to the high-frequency trader working on Wall Street transferring, storing and retrieving vast amounts of data in a split second from a global network of servers.

The popular image of smooth data centres and the illusion of the feathery cloud long worked as a smoke screen, preventing societies from seeing and acknowledging the planetary weight and extractive nature of the infrastructures facilitating our digital obesity. Polished photographs featured in *Wired* magazine, or shared on design blogs, could not convey the colossal noise and unbearable heat inside the server spaces. If the photos would have shown the building's workers, you could have seen them wearing protective headphones to protect them from the loud constant roar produced by the cooling systems. If they could have conveyed the smell, they might have given off the scent of sweat.

Towards the end of the 2010s, many people were shocked to learn that a capital city's data centres used more electricity than all of its households combined, in a time when electricity networks were already overburdened and not enough renewable energy could be generated to fuel increasingly electric societies. Fossil fuels would still be required, continuing to cause environmental damage elsewhere in the world. Next to electricity, drinking water would be used for cooling, while drinking water shortages were looming or already existing in areas with large-scale data. And while data centres proved to be lucrative investments, profits would escape local tax regimes to flow into the pockets of the shareholders and global corporations that had monopolized the storage and mining of data.

The ugly realization of this energy gluttony and corporate greed soon took a central position in the public discourse surrounding data centres. Together with the growing anger over the boxification of pristine landscapes – the seemingly placeless on-demand dream of online shopping turned out to have a drastic spatial impact in the form of abstract XXL distribution and fulfilment centres crowding peri-urban areas and the territories formerly known as countryside – this led to a widespread backlash against data centres.

Era of Entanglement: Data Communities

In the early 2020s, the arrival of so-called hyperscale data centres initially generated a sense of pride among local governments and communities for having
been able to lure Big Tech companies, until concerns about landscape abstraction and energy emergencies became a shared urgency. Cities such as Amsterdam, which had previously auditioned to host data centre clusters, now blocked developments or imposed strict location policies targeting the data centre industry. Recognizing that data centres could not be banned from cities, these policies demanded the urban embedding and multi-functionality of data centres and the reuse of surplus heat to warm surrounding buildings, while restricting the buildings' energy and water consumption.[5] These regulations did spur some technological innovations that somewhat slowed the growing environmental impact through efficiency gains. Green rooftops, fuel cells, thermal energy storage and natural cooling systems, combined with the use of renewable energy, helped to reduce the carbon footprint. Stricter zoning laws also mitigated value extraction of commercial data centre developments, because not every square metre could be financialized: developers were obliged to add public amenities to the privately owned data centres, such as parcel pickup points, urban greenery, cafés and publicly accessible galleries showing private art collections. Nevertheless, many of these 'social' add-ons never felt truly urban, as most data centres remained detached from their urban surroundings. Meanwhile, society's need for data processing and storage exploded through virtual and augmented reality applications, increased AI operations and other forms of expanding digitization.

Real change only happened when local communities organized to build up political clout and gain enough support for the communalization of urban data infrastructure. Following the example of other places that had managed to reverse the privatization of energy production, food and housing, local data cooperatives were established to operate a finely meshed network of locally embedded data centres. If Hamburg could buy back their power plant and grid, other cities could (re)claim their data infrastructure. 6 If Berlin could expropriate 240,000 flats from landlords, communities in other places could recover space for collective benefit. 7 And if Europe's most prestigious architecture prize was awarded to the creation of a co-ownership model in Barcelona, other places could have Good Architecture that designed collective well-being. 8

Amsterdam, Hamburg, Berlin and Barcelona were also among a growing alliance of cities, also including cities outside of Europe including Medellin, Taipei and Jakarta, that had signed the Municipalist Declaration of Local Governments for the Right to Housing and the Right to the City in 2018. 9 They soon expanded their ambitions beyond housing into other community and biodiversity goals. Communal ownership of data infrastructure was one of them, and emerging cooperatives were actively supported by the member municipalities. A typical cooperative would consist of a kaleidoscopic coalition of, among others, horticulturalists, farmers, small business owners and scientists who shared certain economic concerns and interests, together with intellectuals, anarchists and environmental organizations with similar political motives, and various groups with social aims, including senior volunteers who served as community hosts. They organized through monthly Citizen Summits: roundtable discussions through which 100 randomly selected community members would deliberate and decide on issues at hand.

The communalization of data centres coincided with a series of major technological inventions that made it possible to shrink parts of the data centres in size, connect them to other types of infrastructure and further reduce their energy consumption. Data centres slowly shrank in size and started appearing in a wide variety of places, together facilitating the ever-growing amount of data processed. In a growing number of cities they became integrated into public swimming pools, where the residual heat would be reused to keep the water warm. New homes would have an integrated battery system to store solar energy and provide electricity for the neighbourhood's peer-to-peer data storage. When electric cars became the norm, redundant car parks and gas stations were repurposed and developed into hybrid hubs for renewable energy, food production and data storage.

Large-scale data centres would be moved back from remote sites to central locations. There, in the heart of their community, these Data Community Centres

(DCCs) would stand as proud cathedrals of the digital age, collectively conceived and constructed over multiple generations, as a physical reminder of the power of place in times of digital detachment. In some cases, DCCs would occupy and reanimate vacant churches and other redundant spaces of worship and cultural heritage buildings. Barcelona's Supercomputing Center would become one of the first – and most literal – of such data cathedrals. 10 Below ground, in the dark and cold crypts, stacks of servers in sealed-off spaces would facilitate the deep storage of the community's oldest memories. Sunlit chapels would house the docking stations where community members could upload and download their encrypted data. The central nave served as the main data hall, the nerve centre of the network of servers. Rising above it all, the spire, as a potent symbol of connectivity on the city's skyline, doubled as a cooling tower to bring colder air into the building. The vast roofscape was covered with organic solar cells, and the community was tapped into collectively owned networks of hydro- and wind-generated power.

By 2030 all the tech giants had become green, only using renewable energy to power their equipment. However, the criticism towards the emitted carbon and affected ecosystems resulting from the production of the 'green growth' technologies, had by that time increased the understanding that technological fixes were only part of the solution. As cultural labels, 'fossil-free' and 'carbon-free' were soon complemented by 'data-free', to positively signal activities that did not require the production, processing, storage and transfer of data. Besides recognizing data's detrimental effect on the planet, the mental and political well-being of individuals – away from digital mediation and surveillance – became an ambition shared throughout societies.

In data communities worldwide, this focus on post-digital wellness and growth limitations fostered new attitudes. Whenever another megawatt of electricity was generated or a petabyte of data was processed in DCC, a bright light would illuminate the dark sky to remind the community of their excessive consumption and the resource extraction it required. These new insights would inform collective decision-making as to what types of data to store and which digital activities to prioritize within the limits of energy production, storage capacity, carbon budgets and public mental health.

The local grounding of data coincided with Web3 developments such as blockchain technologies that allowed people to own and allocate all of their personal data. All of these cultural and technological developments enabled the decentralization of the Internet and re-politicization of data, and supported the re-humanization of the data centre as civic infrastructure.

Attached to the main building of a typical DCC, smaller structures would emerge, used for public housing and work spaces for social enterprises that develop services to support the democratization of data and energy production and the wider cooperative while cultivating a focus on local intimacy, care and dependency to offset decades of digital globalization. Together, many of these interconnected data communities would come to make up the Era of Entanglement that we find ourselves in today. Looking back, the drastic shifts of the 2020s and 30s have been pivotal to getting here.

1. See W. v. Etten, 'The History of Telecommunications', presentation, University of Twente, Enschede, the Netherlands, date unknown. Online. Available at *www.studieverzameling.utwente.nl/documents/TelecomHistory2.pdf* (accessed 28 September 2022).
2. Quoted in C. Glassman, 'Rare Look at Magnificent Landmark Lobby of Tribeca's 60 Hudson Street', *The Tribeca Trib*, 4 February 2019. Online. Available at *www.tribecatrib.com/content/rare-look-magnificent-landmark-lobby-tribecas-60-hudson-street* (accessed 28 September 2022).
3. New York City Landmarks Preservation, in: M.A. Postal (ed.), *Guide to New York City Landmarks*, 4th ed. (New York: John Wiley & Sons, 2009), 36.
4. 'Skyscraper Begun by Western Union; Largest Telegraph Building in World Will Cover Block in Hudson Street', *New York Times*, 22 August 1928. Online. Available at *www.nytimes.com/1928/08/22/archives/skyscraper-begun-by-western-union-largest-telegraph-building-in.html* (accessed 7 March 2022).
5. See 'Amsterdam Duurzaam Digitaal: Vestigingsbeleid datacenters gemeente Amsterdam 2020–2030', Gemeente Amsterdam, 14 October 2020. Online. Available at *amsterdam.raadsinformatie.nl/document/10115471/1/Raadsbesluit_375_1544A_3a_20_bijl1%20Vestigingsbeleid%20datacentres%20gemeente%20Amsterdam* (accessed 28 September 2022).
6. W.W. Hoeschele, 'How Residents of Hamburg Reclaimed the Power Grid', *Shareable*, 21 June 2017. Online. Available at *www.shareable.net/how-residents-of-hamburg-reclaimed-the-power-grid* (accessed 28 September 2022).
7. E. Solomon, 'Berlin Voters Back Call for Property Expropriation', *Financial Times*, 27 September 2021. Online. Available at *www.ft.com/content/dfabdf27-0d85-4659-9a65-2e84ada7bb0b* (accessed 28 September 2022).
8. S. Burgen, '"Accessible, Not Speculative": Barcelona Housing Co-op Wins Architecture Award', *The Guardian*, 2 May 2022. Online. Available at *www.theguardian.com/world/2022/may/02/barcelona-housing-co-op-la-borda-wins-architecture-award* (accessed 28 September 2022).
9. See the website Cities for Adequate Housing. Online. Available at *www.citiesforhousing.org* (accessed 28 September 2022).
10. See website for Barcelona Supercomputing Center. Online. Available at *www.bsc.es/marenostrum/um&sa=D&source=docs&ust=1655283137444756&usg=AOvVaw2hMwKPjFuJe6iEp4w4R42m* (accessed 28 September 2022).

Liberating Identities – Resisting the Cloud

The fourth chapter 'Liberating Identities' examines from a variety of perspectives the oppression, inequality and exploitation of humans and more-than-humans that results from the imbalanced power relations in the production and control of data and technology. Big Tech is shaped by a capitalist logic of extraction, which, in pursuit of maximizing profit and sustaining the current lifestyles of consumers, does not hesitate to employ methods of neocolonial appropriation and administration. How can we theorize about this or take action to free ourselves from it?

Mél Hogan, director of Canada's Environmental Media Lab (EML) and editor of *The Datafix*, focusses on the computational power enabled by the data centre as a sign of white settler scientific progress and the way it brings us closer to apocalyptic thinking. What kind of human is being created and reflected by AI and computation, and by what mechanisms? Hogan signals that 'in essence, *we become the data centre*: the data centre is of us. AI and humans will share registers of feeling and be components of their shared infrastructure'.

Ramon Amaro's writing, research and practice emerge at the intersections of Black Studies, digital culture, psychosocial studies and the critique of computational reason. In conversation with philosopher/designer **Sheena Calvert**, Amaro discusses how the history of data and statistical analysis informs the complex relationship between racism and machine learning and asks what technology might enable the reconfiguration of Black being. In doing so, Calvert and Amaro anticipate what is to come in the near future.

Focussing on the Atacameño, the indigenous people of the Atacama Desert in Chile, and their struggle against lithium mining – lithium is used to power data centres – and water stress in their salt flats, architect and researcher **Marina Otero Verzier** rethinks the legitimacy of data-driven, extractive societies and their expansion at the expense of indigenous communities. The climate catastrophe makes it necessary to reinvent how data is produced, distributed and stored, says Otero Verzier, emphasizing the need for renewed relationships between digital infrastructures and indigenous peoples.

The Pulse of the Data Centre: AI, Emotions and Planetary Destruction

Mél Hogan

When I began researching and writing about data centres in 2012, I did so because they seemed like a good space to continue my thinking about memory and archives – two themes I'd explored in grad school. After all, data centres are huge warehouses that hold data, so whether as a 'medium' or as a problematic conceptualization of 'the archive', they compelled me. But it was also a perhaps unconscious move away from the *kind* of archival work I'd been doing, which was personal, queerly orientated and emotionally draining because of its social justice components. The data centre, in contrast, seemed affectless – I thought it wouldn't ask too much of me. I was wrong. A decade into this project, the data centre remains an uncomfortable site of research – not affectless as I'd hoped but rather frightening in scale and significance. Research into data centres requires a constant sense of adventure, to get inside, to understand firsthand the spaces that house data, and to grapple with the overwhelming momentum they create for thinking about the future of humans (and humanity) in relation to the data we generate.

Ten years ago, there were news stories about Internet infrastructure, covering both the economic promise of data centres and concerns in how they were replacing other types of industry. But it took almost a decade for these stories to make headlines and ignite public concern about their role in radically transforming political, societal and environmental spheres. Today, it's not uncommon to overhear casual discussions about the political and environmental impacts of technology, of the Internet specifically, while we rely on it more than ever. As with most technology, the discursive unearthing of the data centre serves as a kind of focal point for expressing both societal hype and anxiety through the various meanings everyday people make of it. What does the data centre mean to the masses? How do we imagine the data centre and the role it plays in society? In this sense, the data centre isn't just a repository of data that makes a globally wired planet possible, it's also a social infrastructure that takes on new significance, especially in times of political unrest and climate catastrophe.

In mainstream media and popular representations, more computational power generally gets equated with visions of progress for humanity. With this kind of reasoning, we might think that technology is our best bet for fixing problems, such as salvaging the environment or fighting fascism. But the opposite is proving to be truer. If we counter the idea that the data centre is a kind of hopeful technological object, we see that it enables and exacerbates what is already here and evidences the dramatic splintering between those in control of the industry and the rest of us, subject to its consequences. Specifically, as the imaginary of the Internet moves from already complex forms of communication (social media, financial trading, genomics, space research, etc.,) towards artificial intelligence (AI), machine-learning (ML) and virtual worlds (Metaverse), the data centre itself might be better thought of as a foil, obscuring and obfuscating questions about what it enables, sets in motion and makes global. In this way, focussing on the data centre as a media object or site of inquiry might be like taking the bait from the industry; it entices and distracts researchers from more important humanity- and planetary-scale crises. In data centres, we see the dreams and visions made to matter for the tech billionaire class. As a research site, the data centre requires confronting what has been set in motion.

That being said, in pre-pandemic times, I very much enjoyed my fieldwork in and around data centres – meeting site managers and engineers in the near Arctic, under the constant darkness of the winter season; biking by the Facebook data centre in Luleå in northern Sweden; taking students to a high security data centre in Chicago; driving through Norway with my brother to fields of flowers where the Ballangen data centres were extensively planned but never built, and so on. Over time, it was the many affective dislocations of the data centre that made it most interesting as a site of research: they proposed a solution of sorts – a 'technological fix' – and I wanted to know what problem the data centre had created to resolve.

Unlike my previous research into the environmental impacts of data centres, in this piece, I try to make sense of what it is about humanity that has led us here – to a time and place when we have so much computational power and yet feel so close to the brink of collapse.[1] Of course, as Kim TallBear and others have pointed out, white settler culture is experiencing and documenting the apocalyptic threat now more than ever, while others (many Black and Indigenous people) have long been confronting catastrophes, in what she reframes as a 'sharpening of the already present'[2] (phrasing she borrows from Junot Díaz). With this (re)framing in mind, I focus here on the specific issue of computational power enabled by the data centre as a sign of white settler scientific progress and the way it brings us closer to apocalyptic thinking.

For years I wrote abstractly about the data centre, as though the kinds of data being circulated through the cables and servers didn't matter as much as the overall infrastructural arguments being made. I change course here and focus on one very specific dataset among many, which is how the data centre enables AI and ML, the embodiment of AI, and more specifically, how AI draws from the human emotional register in these planetary conditions. In other words, I look at how the data centre has become the home of human replicas – as holograms, as digital assistants, as chatbots, as deepfakes and so on. All these digital iterations live in the data centre. Without the data centre, embodied AI doesn't 'come to life'. I explore which kind of human, exactly, is created and reflected by AI and computation at a massive scale, and by which mechanisms. I think of what follows as a reflection of sorts; 'taking pulses' along the way. 'Taking pulses' is a method that I offer here for reflecting back on research, trying to get a 'sense of things', or to 'judge the mood' or 'feeling' of a situation. The pulse of the data centre.

Machines Like Us

There are many everyday examples of the deeply limited and most often vile impulses programmed and exacerbated in computer systems at scale, as well as in the tech communities that embrace their own sense of objectivity and neutrality in creating them: I think of Silicon Valley deciding how language should be used (OPT-175B, GPT-3), large-scale investments in surveillance technologies, Google's search engines distorting results based on violent stereotypes,[3] Amazon's recommendation algorithms enhancing conspiracy theories,[4] Microsoft's racist, sexist, raging AI Twitter bot,[5] Alibaba's facial recognition algorithms for racial profiling, disinformation on Facebook working to derail democracy and the misogynist manifesto circulated at Google and their recent firing of AI ethicists among many others. But, as Sarah Sharma writes, 'Sexism at [insert your favourite tech company here] has ceased to be a surprising headline.'[6] We've come to expect it ('it' being misogyny, racism, ableism, etc.) because why would the machines that are programmed by humans not be just like those doing the programming, as well as those that become the datasets from which we teach AI?

Technology was never neutral; the fact that it grows into the likeness of those who conceptualize it seems inevitable. Unsurprisingly, we abuse our likeness in AI form the same way we can, and do, in 'real' life. A recent example of this tendency is Reddit users creating AI 'partners' – in this case, Replika chatbots – explicitly to verbally abuse them, and to make those exchanges available online for others to 'enjoy'. 7

Initially created as 'virtual assistants', the ideas and feelings associated with embodied AI/AI in human avatar forms have already proved violent in the hands of some users. As Dejan Jotanovic writes in *Bitch* magazine: 'The feminization of AI in the form of personal digital assistants is a marker of real women's continued subservience' – and more importantly, he notes: 'She is programmed that way.' 8 So what happens when these machines take over (and by 'machines', here, I mean AI, machine-learning, 5G, holograms, voice-overs, robots, etc.) – not as hostile entities, necessarily, but as unfortunate and mundane continuations of the mysoginist white supremacist ideals always already encoded in the fabric of Big Tech, brought into play by data centres? How will this manifest in a world primed to, for example, accept Amazon as our healthcare providers and Google as our educators? As Toby Ord writes in *The Precipice* (2020): 'Philosophy and fiction often ask us to consider societies that are optimized for some of the things we care about...' and this: '...assumes the builders of the system are striving to align it to human values. But we should expect some developers to be more focused on building systems to achieve other goals, such as winning wars or maximizing profits, perhaps with very little focus on ethical constraints.' 9

We might imagine who sets these goals, and we might worry that those programming the machines to learn new tricks do so foremost for themselves (their bodies, their minds, their anxieties, their visions) – in their own image. 10 There is a component of identity, of power and of obliviousness all baked into this relationship between machine and programmer – and by extension, between programmer and future worlds. Following current technological trends, one world that is imagined is a humanity either replaced by or tethered to the Internet – first by way of our phones, and later by direct signals to and from the brain. AI, ML, brain implants, neural dust, neuromorphic chips and nanorobotics are all evidence of this direction, a convergence between biological and computational systems.

Infrastructure of Feeling (and Reward)

It's significant that our infrastructure is (has always been) a reflection of the limits of our collective political-psychological states: an infrastructure of feeling. The future that Big Tech and the data centre industry are showing us is a psycho-social nightmare of self-commodification, influencer culture, surveillance capitalism, psyops and control-societies.[11]

In anticipation of this very thing – of darker days ahead because of AI's relationship to and encroachment upon humanity[12] – public intellectuals have weighed in: Michael Pollan wonders if psychedelics might be what we need to save us from ourselves, and when asked about survival skills in the next 50 years, Yuval Noah Harari suggests that 'the most important thing is to invest in emotional intelligence and mental balance, because the hardest challenges will be psychological'.[13] And popular culture icons like Oprah Winfrey and Eckhart Tolle propose 'A New Earth' through a series of practical teachings pushing people to reconnect with the sacred,[14] while Joe Rogan and Jordan Peterson invite their millions of followers to deflect and deny the threat of planetary collapse altogether, through 'brain dissolving' banter.[15] These are all settler orientations to the apocalypse, a turn inward. Some settlers may distance themselves from overt climate deniers, but in essence, settler scientific relations to the planet have meant that 'settler' is synonymous with 'denier'.[16] In a gesture of repair, more recent theoretical interventions highlighted by Hi'ilei Julia Kawehipuaakahaopulani Hobart and Tamara Kneese suggest that the idea is to extend 'the potentialities of self-care outward to include other forms of care that push back against structural disadvantage' spurred and exacerbated by neoliberal extractivist capitalism.[17]

Taking this point very seriously, what if these kinds of radical gestures of self-care and emotional intelligence are what's needed to reprogram those programming the machines, and then the AI itself, to recentre and focus emotional intelligence in ML to redefine what is rewarded at a planetary scale?

Perhaps it's obvious from the examples I discuss above that the tech world lacks empathy, and that this has violent repercussions. And perhaps it's obvious too that separating reason from feeling, body from brain and human from habitat, have long served this end. In this way, binaries are reflected in both the code and in the affect of the industry that quantifies emotions and intelligence without truly grappling with either. To understand our planetary futures, we must come to grips with what is being rewarded, how reward becomes the central motivation and the central agent of settler futurities.

Currently, AI works by reward.[18] Much like the simplest take on Pavlov's dog experiments,[19] AI learns by reinforcement: 'learning what to do – how to map situations to actions – so as to maximize a numerical reward signal.'[20] In this learning paradigm, the AI isn't told which actions are correct (that would be the case under *supervised learning*), but rather discovers them through the accumulation of rewards. According to AI specialists Sutton et al., maximizing rewards requires: 'exploration and exploitation', [21] meaning that AI must look to the past to see what has worked, but also forge ahead, into the unknown, to open up new paths and possibilities – for more reward. For many AI researchers, *self-* or *unsupervised learning* is the ultimate goal: 'predicting everything from everything else.' [22] Increasingly, algorithms are being programmed to teach themselves new tricks, what we call 'machine-learning' (ML). ML essentially means that computer programs learn and adapt from large datasets, to generate a kind of AI. While sometimes unpredictable, we invest in ML because of what it promises in terms of its capacity for developing things that the human brain (on its own) can't – more specifically, for its perceived capacity for finding patterns that seem inherent and inevitable; in turn, these serve in making and predicting futures. Deep learning (DL), a subset of ML, is based on artificial neural networks, computing systems inspired by the biological neural networks in the brain. Interestingly, the human brain is already itself conceived as a kind of 'prediction machine', the kind we are trying to refine by ML and AI. But we have to ask – if AI and ML are driven by reward alone – what is said to be worth rewarding? How are reward and emotion related? And what does its opposite – punishment (?) – look like? We can conclude that the reward itself, as a goal for these models, is conceptually problematic. Pragmatically, it's very useful; immediate and self-validating. Either way, DL, ML and AI are the abstracted byproducts of the data centre, the new infrastructures of feeling that shape the Anthropocene / capitalocene.

What does this mean for how we've modelled our global communication infrastructure on our humanity, [23] a bizarre mix of rewarding cruelty and creativity? What do we extract from humans to make infrastructure? How have we remade 'the human' in and through the data centre? And what can we learn about humans from our replicas, built from our own data, as embodied AI? And what happens to embodied AI if / when the data centre collapses?

Emotional Resonance

It's often said that what separates humans from machines is emotional intelligence: empathy, specifically. However, I'd argue that we have all met humans with little to no capacity for empathy. And we have some evidence that AI often has an emotional resonance, if not emotional intelligence per se. And there's evidence that humans function emotionally on the basis of electrical data streams. The line between human and machine is not so clear-cut. For humans, anyway, the difference will soon be next to impossible to decipher. It's likely that within the next five years, humans won't be able to tell real from fake, human from AI, embodied AI from cyborg, etc. But, for now, the fact that we *can* be told apart is likely to bother humans. Ironically, we will soon crave better replicas of ourselves so as to not be put off by the 'uncanny valley': 'a range of appearances, mannerisms, and/or behaviours subtly different from humanoid in an otherwise humanoid figure that may cause negative reactions, such as fear, discomfort, or revulsion.' [24]

Deepfakes – a concept that combines 'deep learning' and 'fake' – is a technology using an evolving form of AI that's adept at making compilations of doctored images and audio into embodied AI (in the most literal sense of the word). This has made it difficult for human senses to distinguish real from fake. This has been vividly illustrated by the 'deepfake Tom Cruise', an extremely popular prototype for AI company Metaphysic. [25] This company specializes in 'hyper-real content' ... that isn't 'real' at all. [26] Again, as this technology became available to customers (around 2017), one of the first uses was to swap celebrities' faces – mostly those of women – with those of porn stars to make non-consensual sexual videos – again, to be shared online by users for others to 'enjoy'. [27] Less overtly misogynist concerns have been raised as well, with issues emerging around the use of materials to bring the dead back to life – whether through animating old photographs, [28] the AI-generated voice of Anthony Bourdain 'narrating' a documentary about his life, or Whitney Houston becoming a hologram for mass entertainment. Each of these has an emotional resonance that is worth unpacking, if only to understand how, through the data centre, we embody AI and give it an emotional resonance. Again, what is revealed by the data centre's infrastructure of feeling?

An obvious parallel can be made here to Raymond Williams's (1978) idea of a 'structure of feeling', [29] which is about our capacity to infer something about a particular trajectory, a movement between thoughts and feelings. In the case of an 'infrastructure of feeling', and especially of the data centre, much of the same can be sensed and said – however, this time with particular attention to the ways in which reward functions.

This infrastructure of feeling is a total departure from the concept of 'The Turing test', which for a long time was key to understanding how human-like machines were becoming by testing for human-like intelligence in machines. The game tested a machine's ability to exhibit intelligent behaviour equivalent to, or indistinguishable from, that of a human. Once a human was deceived by the machine – and the deception here is believing that the machine is in fact a human – then that machine is 'intelligent'. But the test doesn't assess emotion and doesn't situate intelligence within an emotional register. Here again we can see what is rewarded, and what is altogether denied.

As a human being, you can measure and assess your emotional intelligence (or quotient) (your 'EQ') using free tests on the Internet. Just like the traditional (and debunked) intelligence quotient (or IQ), the EQ test was created to quantify and categorize human experiences that are otherwise impossible to measure.

But EQ is really about a kind of emotional calibration that humans themselves have yet to master. A high EQ is usually defined by an awareness of one's and others' emotions, the ability to manage emotions, to tolerate the range of feelings that stem from emoting, and to communicate what we feel as clearly as possible. Doing all of this usually entails reducing our biological stress reactions. Even at a glance, we know these are not values that have laid the foundation for Silicon Valley, Big Tech, start-up culture, or what moves and motivates entrepreneurial thinking in the AI industry (let alone academia or society at large). These industries are about overriding and suppressing emotions through intense workplace competition, and as illustrated earlier, by neglecting the harmful effects of racism, misogyny, etc., made worse by the technologies they develop.

In time, it may be easier to program embodied AI to have higher EQ than humans, since the goal of achieving a high EQ is less about feeling, and more about managing the outcomes of emotionally charged situations. In other words, not letting feelings show, and more importantly, not letting emotions thwart the path to reward. Embodied AI are more likely to parse information for future action because they can't take things personally, and they don't stay wedded to narratives that hurt them, as humans so often do. The outcome of this, of course, could be a combination of the worst of these two worlds: unempathetic embodied AI fuelled by the worst of what the billionaire tech class has enabled
via Big Tech. Regardless, based on these definitions of high EQ, embodied AI are much more likely to succeed here than we are. To be clear, embodied AI will likely have a better EQ than humans specifically because it is not human. It will be interesting to see if embodied AI will host multiple selves, with sometimes competing ideas, like we humans do. Will embodied AI care about the planet as its context for survival? Will it keep secrets? Will their operational algorithms be torn, or indecisive? Will embodied AI sometimes go against its own best interests? Will it lie, when necessary? 30 The main provocation here is that human emotion is arguably only one register of feeling among many. And in the long run, there is a high likelihood of humans merging or hybridizing with AI, or with machines, or with a global communications infrastructure like the Internet. In essence, *we become the data centre*: the data centre is of us. AI and humans will share registers of feeling and be components of their shared infrastructure.

Materially, we must not overlook the fact that AI and ML are now also already part of the data centre, in essence, materializing new politics and emotional registers. We can begin to see how 'the machine' might not need us – AI and ML are used as tools for managing staff, autonomously handling server maintenance, optimization and equipment monitoring, solving complex issues including prediction, assuring business resiliency, optimizing cooling and so on. In this sense, the data centre increasingly relies on AI to run and maintain itself, but it also provides the computational power, space and infrastructure for running AI experiments across the globe. Together, however, these technologies contribute significantly to our growing carbon emissions and the privatization of algorithms. In some cases, AI can help reduce the effects of the climate crisis, such as in smart grid design, developing low-emission infrastructure and modelling climate change predictions. However, estimated costs of training AI models are ultimately at a planetary level, even if they're calculated in dollars. 31

Data Neuroses

Having found their resting place on the innumerable servers that line data centre aisles in countless data centres across the world, the data we've collected tells a story about us. Of late, we've begun converting this mass 'archive' of data into a simulacrum of us. The data centre is at the service of this project. Data centres themselves, connected by vast networks of undersea and above ground cables, constitute a central framework for circulating and perpetuation of a particular way of being and feeling in this world. Seen this way, each new data storage modality brings with it a set of anxieties and responses to those anxieties. But it also brings with it frayed roots and cables. As the material infrastructure breaks down, we are reminded that it – this massive globally networked communications infrastructure we all rely on – was always a short-term project. Undersea cables don't last much longer than a few decades. 32 Hard drives last a fraction of that time, maybe five years. 33 Its time was synced, perhaps, to the end of planetary sustenance. Conceptually, the data centre is part of what Ajay Singh Chaudhary calls 'The Extractive Circuit' when referring to 'an exhausted planet at the end of growth'. 34

Returning to the initial site of inquiry, in light of my critique of their various undergirding impulses, data centres are the massive warehouses that house all of these neuroses. And these neuroses form the archive of the future. Data neurosis is the idea that more data necessarily means better outcomes. Data neurosis also means continuing to invest in broken, violent systems while pretending they serve a purpose beyond self-replication. And in the context of climate catastrophe, data neurosis means having more than sufficient data to make a change – and not acting on it. The data centre makes us complicit with white settler scientific futurities and normalizes its data neurosis.

While neuroscientist David Eagleman proposes that 'the best way to predict the future is to create it', 35 we must also consider just how much the future is predicated on the past and how it works to create modes that preserve, anchor, remember and remind us of the route taken. If rewards are what motivate humans and the machines we teach and train to make autonomous choices, we should carefully consider the implications of immediate rewards. Rewards are not about good decisions for the long-term; they are about instant gratification. They are about the feeling in the moment – perceived risks and benefits. They are about influencer culture 'likes' and rushes of dopamine.

All I can see, from having studied data centres for decades, is that we're setting up a perfect scenario for data to flourish independently from us. It makes it harder to imagine a world without data centres, without technology tracking us and providing feedback that tells us who we are. In a sense the data centre has become a project of tethering humans to wireless devices, and then generating embodied AI from the aggregate data we provided over decades. There's no obvious reason why humans are being created in digital form, as holograms, digital servants and so on – they come into being by way of the vast pools of data accumulated online. In time, they become more sophisticated, more real, more compelling. In time, humans feel more disconnected and displaced from the planet and from one another.

As I was wrapping up this writing project, I came across a new piece of music called 'A Yellow Flower' by artist Henry J. Howe. [36] The musical piece is based on the idea of a sentient machine decoding images in order to understand humanity's relationship with nature. [37] In the music video, we see humans of all ages and backgrounds, close up, mixed with images of butterflies and flowers, seeming to witness their own destruction and demise until it all fades to black. The video plays on the idea that we have the infrastructure – data centres – to provide endless data to train machines, that we use images of ourselves to train machines to feel, and, that humans destroy what is beautiful, and what sustains them – that our emotional register is insufficient to ensure our own survival. I don't know what to conclude here that would be more evocative than 'A Yellow Flower'. Perhaps it – embodied AI, or whatever it will be called by then – will be more easily swayed by collective and long-term rewards than humans are, and better able to ensure the survival of the planet. Only time will tell.

1. M. Hogan, 'The Data Center Industrial Complex', *Saturation* (Durham, NC: Duke University Press). Online. Available at www.doi.org/10.1215/9781478013044-017 (accessed 28 September 2022).
2. K. TallBear, 'A Sharpening of the Already Present: An Indigenous Materialist Reading of Settler Apocalypse 2020', Speaker Series, Department of Political Science, University of Alberta, Edmonton, 10 October 2020. Online. Available at www.youtube.com/watch?v=eO14od9mlTA&t=1512s&ab_channel=DavidJKahane (accessed 28 September 2022).
3. See S.U. Noble, *Algorithms of Oppression: How Search Engines Reinforce Racism* (New York: NYU Press, 2018).
4. See The Disinformation Project, School of Communication at Simon Fraser University, Vancouver. Online. Available at www.sfu.ca/communication/research/projects/the-disinformation-project.html (accessed 28 September 2022).
5. 'The Worst Tweets From Tay, The Millennial Teen AI Bot Turned Racist Nazi', *Funny Or Die*, 24 March 2016. Online. Available at www.funnyordie.com/2016/03/24/77380/the-worst-tweets-from-tay-the-millennial-teen-ai-bot-turned-racist-nazi (accessed 28 September 2022).
6. Sarah Sharma, 'A Manifesto for the Broken Machine', *Camera Obscura* (2020) vol. 104, 35, no. 2, 171.
7. A. Bardhan, 'Men Are Creating AI Girlfriends and Then Verbally Abusing Them', *Futurism*, 18 January 2022. Online. Available at www.futurism.com/chatbot-abuse (accessed 28 September 2022).
8. D. Jotanovic, 'The Future is Fembot: Can We Change the Direction of Gendered AI?', *Bitch Magazine*, 25 April 2019. Online. Available at www.bitchmedia.org/article/gendered-artificial-intelligence (accessed 28 September 2022).
9. T. Ord, *The Precipice: Existential Risk and the Future of Humanity* (New York: Hachette Books, 2020), 144.
10. See A. Taylor, 'The Automation Charade', *Logic* (August 2018), no. 2. Online. Available at www.logicmag.io/failure/the-automation-charade (accessed 28 September 2022).
11. I. Kottasova, 'Elon Musk's Next Move: Merging Brains and Computers?', *CNN*, 28 March 2017. Online. Available at money.cnn.com/2017/03/28/technology/musk-neuralink-brains-computers-ai/index.html (accessed 28 September 2022).
12. Elon Musk et al., 'Superintelligence: Science or Fiction?', *Beneficial AI Conference*, Future of Life Institute, Pacific Grove, CA, 7 January 2017. Online. Available at www.youtu.be/h0962biiZa4 (accessed 28 September 2022).
13. C. Skipper, 'The Most Important Survival Skill for the Next 50 Years Isn't What You Think', *GQ*, 30 September 2018. Online. Available at www.gq.com/story/yuval-noah-harari-tech-future-survival (accessed 28 September 2022).
14. 'Awakening to Your Life's Purpose', *Oprah*, season one, episode 104, chapter 4. Online. Available at www.oprah.com/app/a-new-earth.html#! (accessed 28 September 2022).
15. J. Crosbie, 'Joe Rogan and Jordan Peterson Wax Idiotic on Climate Change and What It Means to Be Black', *Rolling Stone*, 26 January 2022. Online. Available at www.rollingstone.com/culture/culture-news/jordan-peterson-joe-rogan-interview-climate-change-1290696 (accessed 28 September 2022) and M. Taft, 'Jordan Peterson and Joe Rogan Talking About Climate Change Will Make Your Brain Dissolve', *GIZMODO*, 26 January 2022. Online. Available at www.gizmodo.com/jordan-peterson-joe-rogan-climate-denial-1848425540 (accessed 28 September 2022).
16. With co-author S. Cahill [J. Allen et al., 'Un-/Learning Archives in the Age of the Sixth Extinction', workshop, Whole Life Academy, 15 March 2022. Online. Available at wholelife.hkw.de/un-learning-archives-in-the-age-of-the-sixth-extinction (accessed 28 September 2022)] we are writing a series of reflections on the question and concept of denial.
17. H.J. Kawehipuaakahaopulani Hobart and T. Kneese, 'Radical Care: Survival Strategies for Uncertain Times', *Social Text* (2020) vol. 38, no. 1, 1–16. Online. Available at uwethicsofcare.gws.wisc.edu/wp-content/uploads/2020/04/Hobart-and-Tamara-Kneese-2020.pdf (accessed 28 September 2022).
18. The idea of 'reward' is key to understanding how the future is imagined in these computational systems. Reward mechanisms are what most closely link human and machine incentives. The outputs are either by reinforcement learning, or one of the other paradigms for machine learning: supervised or unsupervised learning.
19. M. Specter, 'Drool: Ivan Pavlov's Real Quest', *The New Yorker*, 17 November 2014. Online. Available at www.newyorker.com/magazine/2014/11/24/drool (accessed 28 September 2022).
20. R.S. Sutton and A.G. Barto, *Reinforcement Learning: An Introduction*, 2nd ed. (Cambridge, MA: The MIT Press, 2018), 1.
21. Ibid., 3.
22. R. Toews, 'The Next Generation of Artificial Intelligence', *Forbes*, 12 October 2020. Online. Available at www.forbes.com/sites/robtoews/2020/10/12/the-next-generation-of-artificial-intelligence/?sh=74b2fb1f59eb (accessed 28 September 2022).
23. See T. Adorno, *Negative Dialectics* (New York: Continuum, 2003).
24. See Soul Machines, 'Introducing Soul Machines Human OS 2.0. The Next Generation of Digital People Powered by Autonomous Animation Are Here'. Online. Available at www.youtube.com/watch?v=rjclQ3m5JRw&ab_channel=SoulMachines (accessed 28 September 2022) and A. Ong, 'This Company Is Making Digital Humans to Serve the Metaverse: Meet Sam', *The Verge*, 27 October 2021. Online. Available at www.theverge.com/2021/10/27/22746679/soul-machines-metaverse-digital-humans-labor; www.wordnik.com/words/uncanny%20valley (accessed 28 September 2022).
25. See Metaphysic. Online. Available at www.metaphysic.ai (accessed 28 September 2022).

26. R. Metz, 'How a Deepfake Tom Cruise on TikTok Turned into a Very Real AI Company', *CNN Business*, 6 August 2021. Online. Available at *www.cnn.com/2021/08/06/tech/tom-cruise-deepfake-tiktok-company/index.html* (accessed 28 September 2022).
27. R. Metz, 'The Number of Deepfake Videos Online Is Spiking. Most Are Porn', *CNN Business*, 7 October 2019. Online. Available at *www.cnn.com/2019/10/07/tech/deepfake-videos-increase/index.html* (accessed 28 September 2022).
28. MyHeritage. Online. Available at *www.myheritage.com/deep-nostalgia* (accessed 28 September 2022).
29. R. Williams, *Marxism and Literature (Marxist Introductions)* (Oxford: Oxford University Press, 1978), 126.
30. Drawing from David Eagleman's and Lisa Feldman Barrett's insights, *truly* programming AI to feel like humans would mean working around these very human truths: that emotions can't be detected through facial expressions (or physiology generally); emotions are cultural rather than universal; emotions are not located in specific parts of the brain; and emotions are not reactions to the outside world. 'How Emotions Are Made: The Theory of Constructed Emotion', Forte Labs, post (20 January 2022). Online. Available at *www.fortelabs.co*/blog/how-emotions-are-made (accessed 28 September 2022).
31. P. Dhar, 'The Carbon Impact of Artificial Intelligence', *Nature Machine Intelligence* (August 2020) vol. 2, 423–425. Online. Available at *www.nature.com/articles/s42256-020-0219-9* (accessed 28 September 2022).
32. D.W. Brown, '10 Facts About the Internet's Undersea Cables', *Mental Floss*, 12 November 2015. Online. Available at *www.mentalfloss.com/article/60150/10-facts-about-internets-undersea-cables* (accessed 28 September 2022).
33. 'How Long Do Hard Drives Last? Lifespan and Signs of Failure', Prosoft Engineering, post (15 September 2020). Online. Available at *www.prosofteng.com/blog/how-long-do-hard-drives-last* (accessed 28 September 2022).
34. A. Singh Chaudhary, 'The Extractive Circuit', *The Baffler* (November 2021), no. 60. Online. Available at *www.thebaffler.com/salvos/the-extractive-circuit-singh-chaudhary* (accessed 28 September 2022).
35. David Eagleman, *Livewired: The Inside Story of the Ever-Changing Brain* (New York: Pantheon, 2020), 79.
36. HOWE, 'A Yellow Flower', *Forever*, EP, Higglers Records, 2022. Online. Available at *www.youtube.com/watch?v=cM-BaR-9Vnw&ab_channel=FACTmagazine* (accessed 28 September 2022).
37. H. Bruce-Jones, 'Howe Finds Human Emotion Within Machine Intelligence in "A Yellow Flower"', *Fact Magazine*, 19 January 2022, Online. Available at *www.factmag.com/2022/01/19/howe-a-yellow-flower/* (accessed 28 September 2022).

On the Edifice of Data Centres and the Logics of Knowledge Capture

Ramon Amaro in conversation with Sheena Calvert [1]

138 Sheena Calvert: Data posed as an 'adversary' appears in the language of Niels Schrader's contribution to *Acid Clouds*. [2] You have spoken elsewhere about humanity's historical quest for knowledge, and how this ultimately creates a material residue of that quest, in the form of physical artefacts (books and records, for example). [3] In what ways are data centres different or the same in terms of materializing such searching? Do all repositories (material or immaterial), produce what you call 'aggregated and accumulated data', and in the process of doing so become infrastructures of power? In what ways are these data centres more malign power/informational systems than any other forms that predate them?

Ramon Amaro: In many ways, your question speaks to the tensions that are played out in contemporary data-driven spaces. But it also speaks to the disparity – or distance – between public discourse, our public anxieties with data, and the seeming indifference to these anxieties on the part of those who have accumulated a disproportionate amount of power collecting, processing or using data in the public domain. These, of course, are the individuals involved in the creation of data infrastructures as well as the financial structures around them that act with impunity. The public, nonetheless, approaches these repositories at a distance from any general understanding. Few know what data are or how they are used for financial trade, let alone what kinds of data are captured from their persons or behaviours. Most of us feel powerless at the thought of this massive industry. Data are atmospheric in that they are difficult to place or understand, or even reconcile for the average person. There is an unpleasantness to this mystery, especially when the circulation of power and control remains obscure. At the same time that data circulate through advertising and monetary exchange, they produce ways of knowing the world and influence what the general public had considered to be knowledge. It is within reason then that we would see data and its infrastructures and processes as adversaries.

Sheena Calvert: Talking of the 'real', I understand that 90 per cent of money is just on spreadsheets and in the form of data. We have this mythical idea of money as being physical. I took out money today to pay for something and it was extraordinary to have a physical note in my hand. And yet it isn't real. The real has moved elsewhere. Data is the 'real', if we can still use that word.

Ramon Amaro: Data is very real, and so is physical currency. Yet, at the same time, they are invisible flows of information about what we value in society. Data in particular are flows of information that are invisible to the majority of the human population. At the same time they speak to the utter disregard data has for human participation, or I should say a participation other than one of pure behavioural extraction. Under the premise of farming data in such
quests to achieve greater knowledge, we discover that – perhaps pulling away from early cybernetic fascinations with animal and environmental study – we are now in fact farmed and treated as the fragmented beings to be studied, correlated, conquered and controlled. This is underwritten by a wider logic of social engineering through an ideological framework that knowledge is the exclusive domain of a few who have taken it upon themselves to define what knowledge is useful, what knowledge should be gained and ultimately what knowledge should be held onto or buried under the guise of usefulness. The obscurity of these processes misplaces or, I should say, distracts us from the real cause of our anxiety, namely the uncomfortable truth that the logics of positivism endure in the material and immaterial artefacts of data, and their supposed importance in the aimless search for knowledge. In other words, focussing on the data in and of themselves is a distraction from repeatable – and perhaps hidden – operations that even unwittingly mediate the flow of ownership over knowledge.

Sheena Calvert: Where are the various 'spaces' (both physical and virtual), in which we might find these 'operations'? Where do you think we should be looking for evidence of them? Data centres are one example from the physical world, but what others might we look towards – ones that similarly mediate ownership of knowledge?

Ramon Amaro: These spaces live in our collective knowledge or amnesia of the past. We've seen this before by means of religion, politics and education (to name just a few). They are ghosts from a past project, posing as ideal models for a future society. What we see now in the data are operations that are epistemically tangible and yet intangible and concerned with the ownership and survival of a power over data. Meanwhile, we are left with a circulation of confusion where fact is conflated with fiction, and the past with the future at the expense of the present. This is underpinned by promises of social equitability and objectivity, which we now know is short for institutional control and survival. So, if these physical data centres are anything, they are monuments to institutional and corporate survival, as temporally significant as they are spatially terrifying. And, if we like anything in the West, we like a monument. We love to manifest our anxieties about the future in physical architecture that symbolize a prophetic process of past success or failure. We hope that these monuments help us compile fragments from the past that might help us reconstitute our disjointed relationships with each other. But instead, these monuments, whether they be a statue of a general or a building full of data servers, serve as mere reminders of a world never fully realized yet still longed for in the image of a utopian ideal.

Another way that we can look at it is that these monuments were never about the past, but an opportunity to keep open the possibility of the future: to control the future. Take colonialism, for instance, and how the monument stands in for the violences that fragmented entire communities and communal relationships. We could say that even a colonial street sign, as monument, is a material manifestation of a historical process, seen in the present, something that attaches itself to the celebration of this or that historical type of ideology. We're used to seeing these monuments as being an image of the audacity of homage to something that was deemed to be ideal even though those moments only benefitted a few. We're not used to seeing these monuments as a warning of a future to come if we're not careful.

Sheena Calvert: The colonization (and destruction) of the Mayan cultures, for example, involved destroying any pre-existing knowledge systems.

Ramon Amaro: Absolutely. And this idea is not only terrestrial but interplanetary. The thought is: I go to the moon, but the journey itself isn't enough of a learning experience about the self, about others or about the wider environment. It has nothing to teach me other than that which I prescribe as worth knowing. Otherwise, I must destroy or supplant the knowledge systems that I view as having little utility for myself, or those that support my enterprise. I am the arbiter of the universe's mysteries. I stand above it in a place of seeming objectivity, even though I, myself, am amongst the properties I seek to discover.[4] This narrative is as old as the hero's story, except there is no character arc or transformation at the end, only the drive for aimless innovation and discovery. So just as I might plant my flag on the moon as a symbolic marker of dominance over the flow of cosmic information, I might, within reason, see justification to construct other monumental symbols to demonstrate my Promethean powers for mediation. For instance, look at the physical infrastructure of data centres here on Earth, comprised of sophisticated physical materials constructed in the image of existing architectural form, that – like a national flag on the moon – are rudimentary in representation compared to the sophistication and wonders of any form of data they could possibly designate.[5]

The spaces of colonialism are in no way novel. These spaces also exist in theory and philosophy. There's a long, philosophical history of concern with the so-called nature of lived experience, where many have grappled with the humanist quest for knowledge, and not *only* our quest for knowledge, which sounds quite general, but the issue of our agency within this knowledge quest. We can even go back to the previous Nietzschean type of debates, concerning our place in this world, our positionality, and how much agency we have. Including what is our will, what is determined and what is indeterminate. We can also ask what knowledge has become available to us, what does it mean to capture those knowledges? We've assigned so many forms of value based on that logic, including privileging the erasure of cultural knowledge and supplanting new knowledge systems as an act of discovery.

So, to return to the question, we could say that the data centre is a physical representation of these kinds of power dynamics, even though they function differently. The problem with data is not that it's retrospective.[6] The other problem with data is that their value is based on utility and not an aspiration to because we don't aspire to be anything other than the replication of what we've always been. We live in a contemporary environment where even our claims to an identity are based upon past mechanisms. Everything has (temporally), collapsed. Present interventions, the possibilities of a future, are just reinforcements of the past. We're locked in, and anything that's considered hyperstitious,[7] and by this I mean (prophetic, or even that which might appear aspirational), can be perceived as fantastic, unimaginable or utterly untenable. We've written ourselves out of any future. It's not that the AI might come and kill us – the thing is, I think we've already killed ourselves. And by this I mean our unrealized futures in favour of what we *imagine* ourselves to be.

Sheena Calvert: I was thinking about these data centres as being almost like books that contain the language, and that language, in turn, codifies the knowledge.

There's this assumption that seeking understanding requires us to capture something and bring it into a form where it can then be disseminated to others. Like a book, a data centre is an edifice and a structure that contains knowledge that has been stripped out of a process. We assume that this process is a good one and that what we should do is embrace our nature as meaning-seeking animals, seeking understanding and knowledge, even though it may not be necessary that we do that. We're constantly trying to bring that process down to ground, within a material space, whether that's a data centre, a book or 'other'. But actually, when you're reading a book, you're looking at the flows of language. When you're looking at a data centre, what you're looking at is not the actual building itself, you're looking at the flows of information, and the storage and the materialization of those flows, within the physical space.

Ramon Amaro: What I speak of is very much about language: those structures of language that become carriers of our perceptions of ourselves and the world around us. This includes the view that the world exists – in obligation – for the purposes of our own insights and understanding. It comes along with this quest for discovery, as if the infinite phenomena of the world (the infinite possibilities of the world), are something that we can capture, something that we can observe. It lives alongside a general type of audacity that we have as human beings (not in a sort of cynical way), but as an audacious stance, which is the positivist stance. And by positivist, I mean the individual, company, policy or process that emphasizes observable facts and excludes much of anything else.

Nonetheless, language is a carrier at the top of a wave, through which currents of knowledge circulate. It aligns with the idea that discovery is best served externally rather than within. This is the *audacity*. We adopt the idea that it's not our job to *experience* the infinite phenomena of the world, it's our job to *understand* it, and if I don't understand it, then it's my job to do everything within my capability and power to make it coherent and to make it understandable, to ease our own anxiety. So instead of us dealing with our anxiety, instead of us dealing with *us*, we continue this process as a functional mechanism within the world. And so, when we say, where's this power placed? It's placed within every single one of us. It's just that some of us are more lucky or unlucky than others to make these languages actionable in the form of a start-up, a dune-shaped data centre, an advertising model or within new modes of resistance.

Our preferred language today is data. It is within reason, then, that we would create repositories for the storage and processing of our bounty of supposed knowledge. We are a species obsessed with data and mathematics as the carriers of these information flows. At the same time, and despite our crowning of data and maths as objective truths, we still only want to see the ones that we want to see. Because to do otherwise would require us, collectively, to renegotiate our attachments to meaning, and to acknowledge the sticky consequences of the messages we (even unintentionally), send out into the world, which are of course then replicated by algorithmic mimics such as AI.

Sheena Calvert: There's this idea of the kind of clean, sterilized and neutral façade of the data centre as also being environmentally neutral. There's a lot in *Acid Clouds* about the problem of data mining, and how it produces negative material effects on the Earth. I'm currently reading Kate Crawford's *Atlas of AI* (2022), in which she talks about this fiction of neutrality. What is at stake in presenting these data centres and data itself as neutral, immaterial, metaphysical, beyond the physical world? It's an old process – one allied with capitalism. That's what capital does, it removes you from the consequences of your behaviours and actions. So, the question is: What's at stake and who is the 'they' behind this clean, sterilized façade? Should we be worried about this sterilizing of the truth about what goes on inside these data centres? Or, is it just a distraction?

Ramon Amaro: Let me start by saying that in many ways it's within reason that we would presume or even desire a type of objective datafication or data process, one that has freed itself in terms of ethics, freed itself of any impunity, freed itself from any subjective claim, claiming instead: 'It's outside me.' But it's interesting that this pursuit adopts the language of neutrality, which itself has become a type of Western audacity, and which is coming directly from humanism and the Enlightenment.

This is the Kantian nightmare that theorist and artist Denise Ferreira da Silva and many other radical theorists are trying desperately to get out of. The fundamental logic of Immanuel Kant's *The Critique of Judgment* consists in having this capability to adopt one's own reason and consider all the other phenomena in the world to be sub-processes or secondary processes to that reason, which (in turn), will allow me to access/channel a divine sense of being, as misguided as this sounds. 8

This type of reason is misguided, as it promotes the possibility of decision making, whereby objectivity takes precedence as one that is considered good and reasonable. What's interesting is how quick we are to then adopt the languages of science and quantification, these epistemic languages that emerged from the Kantian moment. Then, what follows is that we adopt this idea, so that we can create a controlled environment to test the possibilities, under the auspices of objectivity and equitability. It's very interesting, since it allows us to create the questions based on the answers we want, thereby controlling the experiment under the auspices of an enlightened and achievable utopia.

Sheena Calvert: We've written ourselves out. The world is going to end because ChatGPT is in it. It's because we're not going behind the surface and looking at it. Meanwhile, what you've just pointed out is it's not just the 'they', as in the big corporations who create a situation where we are simultaneously anxious about the technological future, while at the same time being encouraged to embrace it, unchallenged. We do this to ourselves, unprompted.

Ramon Amaro: Actually, we are the 'they', the people that make up corporations as well as the people that reflect corporate values, even the ones contrary to our own well-being. But from a psychosocial point of view, it's revealing in and of itself. At the risk of sounding like a technical solutionist (because I'm not that way at all), these anxieties can be used in a more value-added sense to promote more affirmative ways of living and being. And by this, I don't mean we should use large language and generative AI models, like ChatGPT, to do our bidding, but we can use its stated results and our anxieties with them to signal an incompatibility between how we perceive ourselves, which is often at a distance from reality, and how we truly want to be as individual and collective human beings.

We've created a technically mediated society whereby any individual or collective with enough technical knowledge or venture capital investment is given

full license to exercise power or restraint. This applies to anything or anyone that attempts to break the fantasy of a more efficient, optimized world wherein its mysteries must not only be comprehended in order to be given value, but also that this value is driven by a misplaced belief in the power of data to solve all of our problems. If anything, what we have written out is any unmitigated reflection on the real symptoms of our human condition that perpetuate and protect violences such as racism, misogyny, homophobia, transphobia and the like, which is well rehearsed through our own historical tendency to treat the other as stranger. Yet we look upon AI as an adversary, for instance, as if its regurgitation of these violences is a crisis in itself, as if it can have success in value-added living where we have failed, as if it can bring us closer to solving the uneasiness that we refuse to see, abolish or reconcile in ourselves, or as if AI – this mirror held up to our human condition – is our last chance at being a final saviour from these dystopian fantasies. Then, if it doesn't behave as we expect, we chastise the algorithm as if it dared to show us in full view the violence we have become.

But there is another choice we can make in how we view this relationship. This could happen if we were to open ourselves to something more generous and aspirational, something other than that which we've become conditioned to, which is in most cases fear and anxiety. What if this alienation we both feel and desire through the technological was to open up to the possibility of new ways of being and relating, including new types of decision making? What if AI were allowed to show us a new way to relate, outside of our present-historical restrictions, towards new possibilities of being that might just show us a way forward without leaving anyone behind?

Sheena Calvert: That's a beautiful thought.

Ramon Amaro: I believe it is a necessary thought, although I should say it is not nuanced techno solutionism. If there is an urgency with AI, then I wonder what generative, value-added, possibilities might be revealed by testing it on an equitable future (one not yet realized), as opposed to optimizing it to moderate and mute the violences we for the most part refuse to face.

Sheena Calvert: Returning to the issue of data centres and coloniality, are you suggesting that they replicate the logics of coloniality in the ways in which they operate through data? What (exactly), are these power relations, and in what ways are they performing or enabling colonialism? I just would really like to pull out and unpack this term, 'digital coloniality', because it's connected, obviously, to what we're talking about here, and about the logics of power and racism, and all the things that exclude and marginalize, and destroy and end in violence.

Ramon Amaro: Thank you for that question because this issue of coloniality in general exists as part of the complication of what we've been discussing here. I've been working with this idea of coloniality, to pick apart the logics of the colonial process itself. First, it's necessary to separate the act of colonialism from the logics of colonialism, and then, of course, relate colonialism to imperialism, which we know is an ideological framework that in many instances remains hidden behind the logics of capital exchange.

What is increasingly important to me, is the role of the individual and their relation to the collective, but not as a way to replicate rhetoric or philosophies of individual will or collective possibility, or to obsess about agency or place, or attempt a reconciliation of being. What I'm thinking about instead, is this act of thought at the point of decision making, whether these decisions are derived from human, technical objects or a combination of both. The focus of my work at this point is thinking about who and what's *left behind*, and to really think about these logics of colonialism entails not just what conditions allow colonial processes to emerge, but what conditions allow the individuals and collectives to perpetuate those logics, to be indifferent to those logics or resist them altogether. Ultimately, the concern is how we release this responsibility onto 'the mathematical' (or the algorithmic) as a proxy for what in some cases serves as a very human tendency to reduce the world to symbolic form, and thereby dehumanize everyone and everything in it. Returning to the colonial then, what better way to surrender humanity over to violence? What better way to reduce responsibility and accountability for our human condition than to hand over the act of dehumanization to that which is non-human, which in history has flowed seamlessly through mathematics, data and other quantified information?

Sheena Calvert: This is of course what Theodor W. Adorno, in *Negative Dialectics* talks about when he speaks about the administrative murder of millions, the ability to render human beings as merely numerical data within a system, such that they just become non-human numbers, and therefore they're not real and then you can kill them.[9] One of the key logics of this kind of coloniality and racism is in the removal of any human contact/proximity, so that you render someone as simply a number within a system. That logic and process exists before we get to data and their centres. These same processes and logics have become memorialized and materialized in these physical buildings that we see on the street. Adorno reminds us that it's really these fundamental relations that we need to think through, before we get to any instantiations of them (in data centres or 'other').

Ramon Amaro: For me, it is important to return to a Fanonian moment of coloniality where violence is not merely an epistemic act, but symptomatic of a larger social diagnostic concern.[10] This is why I find it necessary to return to the psychosocial. In order to truly understand the logics of colonial violence, including its contemporary manifestations in data or acts otherwise, it is important for me to expose the individual – even my own individual sense of being – and collective body, to our apparent, and even unwitting, awareness of our psychic flows of thought that inform our acts of decision. In other words, I seek to understand how one arrives at a moment of violence, value or indifference. Because, let's be real, each moment is in flux and might appear to be single acts but these acts are more accurately articulations of an amplified set of conditions that are as historically informed as they are presently urgent. This historico-present information takes active work to maintain, even if we are not aware that we are doing so. What, epistemically, draws us to carry this awareness out onto the surfaces of our individual and collective psychic awareness? And most importantly, how can this, what I like to call *radical awareness*, increase our chances for catalysing new, more value-added, thoughts and decisions.

Sheena Calvert: This reinforces that we can't just throw our hands up and go into a state of paralysis. Coming back to a related point, one of the ideas that comes up around these data centres is how data has been clean and sterilized. We might say that sanitization is part of the process of distancing, of the removal of responsibility. It wasn't me. I didn't do it. Somebody else did. We present this thing called the data centre as 'the cloud', employing a benign metaphor, an old-fashioned marketing trick. Similarly, 'collateral damage' is the phrase used for the murder of civilians who had nothing to do with the war but are casualties of that war. Such metaphors take attention away from the damaging consequences of our actions, and we fall for it every time. How do we counter such misrepresentations and deflections away from what's really going on? How do we expose these layers of deceit, manipulation, coercion, exploitation and violence against people, against the body politic, against democracy?

Ramon Amaro: Precisely. What I seek is action, whether they be habitual, planned or inspired. But not just physical actions, psychosocial action is paramount. What good would it do us to hand this difficult process and any of its future possibilities over to data? Instead, how can we reshape our psychic proximity to our thoughts and experiences in the world, despite its duress? How can we create a world where these acts – even those mediated by data – open pathways for us to take command of our own psychic potential to be, as humans, better than what we've historically experienced? In other words, if we are going to commit to maintaining these reference books, as you described them, called data centres, then why can't they instead function as repositories for alternative thought and methodologies for more affirmative and communal relationships in the world?

Sheena Calvert: Meanwhile, as Adorno remarks in *Negative Dialectics*, but within a very different context, drowning out screams is a mark of the complete loss of humanity. 11

Ramon Amaro: Until we recognize the duress and the historico-present logics that substantiate violence, fragmentation and dehumanization – until we face the difficult task of our own habitual contributions head on – with a radical type of actionable awareness, as an everyday psychosocial practice in itself, then we will continue to drown out those screams, and even more dangerously blame them on algorithms, as if they are the cause of this nightmare.

Sheena Calvert: Then you've got philosophers like myself who are asking: what is language in this context? I'm continually coming back to that question.

Ramon Amaro: I don't want to distract from the utter importance of language in this sense, and of course we are talking about language in different registers here. But perhaps the answer is in the anxiety of the question. Perhaps that is what helps philosophers such as yourself stay with the trouble, as Donna J. Haraway would say, 12 but keeping you within the process as opposed to the impossibility of a solution. I would go as far as saying this is what models such as ChatGPT get wrong. These large language models are programmed to assume that the dynamic flows of language in all of their manifestations are situated knowledges that demand simplified and coherent solutions, whereby philosophers such as yourself see language as instead.

For me, the need to give language over to coherence is an exhausting process, as it demands that I also give over my capability to aspire to or imagine new possibilities to restructure and rearticulate the languages that already exist. And I don't want to stay here within the stasis of unrealized language and human potential. I don't want my existence, or the languages used to express it to be measured, correlated and parroted back to me for efficiency. I desire for language I speak in all of its capacities to remain in view as something that indicates the wonderous impossibility of being able to contain the overwhelming expansion of living and being in itself. If anything, I aspire towards less language in this sense, which has historically closed off human potential as we've seen in our technological, not to mention political milieu.

Sheena Calvert: I love the way you're saying this, but I'm going to be challenging. Are you knowingly presenting a sort of technological utopianism that's naive given the power and the speed at which these things are moving?

 Ramon Amaro: I am what I struggle to describe as 'future' naive.

Sheena Calvert: I love that you would admit that.

Ramon Amaro: [laughs] I'm not shy in saying that I have grown to realize that the more I know and attempt to know, the more I recognize that which I don't know. And I enjoy this process. It keeps me in close proximity to the limitations of my understanding, but it more so helps me connect to an expansive universe of potential that would be impossible, as humans, to fully comprehend in scope or possibility. I can say this openly because naivety for me isn't a bad thing because it presupposes and reinforces – even in the sense of my own individual being – the concept that the knowledges I acquire are directly proportional to the infinite possibilities of being in itself. I haven't even cracked the surface of what is possible. And this forces that naiveté into a space becoming that – in my view – lends itself to a more authentic way of being, one that demands constant reconciliation between how I am presently and historically and what and whom I want to be. This type of naiveté breaks my illusions and sticky perceptions about the world, and forces me instead, to live within the process of transforming between one state or perception of being into another that can't be pre-empted any more than it can be fully comprehended. This is the motivation behind my book, *The Black Technical Object, Machine Learning and Aspiration of Black Being*. 13

In the book, I advocate for reflection on the work of psychiatrist and political philosopher Frantz Fanon, Jamaican writer and philosopher Sylvia Winter, and philosopher of 'individuation' Gilbert Simondon to help make sense of our enduring relationship between technology and human difference. I wanted to explore whether our current distance from human and technical agreement, which as we know is replicated in algorithms, can be located in our historical relationships between ourselves and the world, even prior to what we know today as algorithms. I ask what other 'books', as you say, substances and repositories of knowledge have staged the conditions for our present anxieties that are in turn disrupting our future aspirations. What I discovered is that our tendency to view technology as a stranger had been already rehearsed in our technologically mediated view of other humans as strangers, as compared to the Western symbol of Man [*sic*]. What I mean – following Simondon, and going back to your previous comment about dehumanization – is that even prior to the predominance of colonialism, humans had already become accustomed to viewing the racialized Other as a stranger to humanity. These ideas were, of course, logically reinforced by philosophy, science and medicine to help construct an objective basis for these racist views. As populations grew, and each socially altering technology emerged, so too did the sophistication of mathematical techniques used to reinforce these same logics, under the premise of objective truth. The relationship between technology and humans, and between humans and other humans, is more parasitic than oppositional, yet the basis for control has always been nested within the ability to make use of technology to perpetuate ideas of difference and hierarchy, or the justified stranger. Technology has less so been put into use for the purposes of reducing this divide, which would require an imaginary that exceeds our present understanding of how various entities relate in and to the world, whether they be human or non-human. Yet any anxiety that transposes can be viewed as a signal of and to the incompatibility between our imagined selves and the real truth of the world we created in partnership with technology. The anxiety is a fire alarm that alerts us to the need to transform into this relationship and ourselves into something otherwise.

Sheena Calvert: In other words, we need to be in a state where we are staying with the trouble?[14] We need to stay with this trouble, since the struggle is productive.

Ramon Amaro: You make use of the anger. You make use of those anxieties. You make use of what Simondon is saying, of the incompatibility of the relation in itself. This is where Fanon was going. You make use of the incompatibility to aspire, that's formed through the urgency to create a new moment of being the unexpected and historically unrealized.

Sheena Calvert: In other words, what you appear to be saying is that you write the contradiction. You retain the space for that beautiful old philosophical word, 'incompossibility', that which is incompatible, but still has the possible within it.

Ramon Amaro: And that's what I write about in the book with reference to Gottfried Wilhelm Leibniz, the seventeenth-century philosopher of the monadology. 15 But you can't think of it in the monadic sense, because that incompossibility would then presume that there's a single origin from which all difference emerges. However, you set that free, which is what Gilles Deleuze and Félix Guattari tried to do, and if you free it from that (limited) monadic sense into something else, then you have an infinite number of possibilities; an infinite proliferation made possible through a different model of thinking.

Sheena Calvert: Absolutely. There's something beautiful in the notion of all these lines of possibility. Deleuze and Guattari asked: how do we capture these lines of flight? 16 This beautiful way of thinking, this beautiful notion of the lines of flight, and the movement that allows for other things to emerge, other things to happen. It's not just theoretical. It's a proposition for an actual process that we can live our lives through.

Ramon Amaro: I'm using some of this method as a shortcut towards the disruption of Whiteness, but ultimately the physical, systemic and psychic liberation of Black being. And I'm not talking about a race of white people, I'm talking about the concept, the ideology of Whiteness as the repositors of Truth and the distribution of value and meaning. The West has to give up on this idea that equitability means equal at all times, because for communities outside of Whiteness, equitability is an exchange of care, which means that sometimes you are giving more than you're receiving and at other times you're receiving more than you're giving.

Sheena Calvert: Margaret Bowden, who's written about AI, 17 has said we don't understand how the human mind works, so we can't really replicate it in any machine systems. But she has also said that AI asks us to think about very old questions about who we are, what we are, how we are, how we think about power, about, all the things we're talking about today. And at its very minimum, even if we don't have true artificial intelligence, what we do have is an opportunity through it to reflect on these other bigger philosophical questions. AI is forcing us to come back to them. And so, I'm looking at these data centres, and realizing – as you mentioned before – that it's not really about the data centres. The conversation is about what they enable in terms of a discourse around these questions and more. Going full circle, hasn't that been the tension in the past few years with these other types of colonial monuments?

Ramon Amaro: We've allowed them over the years to become these passive relics, and yet you've had these anti-racism, anti-colonial movements, and the general public, who are demanding that if this is going to be a marker – because the edict, 18 was clear – if this is standing as a marker, then it needs to force the conversation. If not, it needs to go. So, the conversation is a lightning rod of utility with no aim. And if there is an aim, it is not used for the purposed of better inter- or intra-human relations. The good news is that this absence can serve as a signal towards the desperate need to reposition technology towards something otherwise.

At a community level, this information can alert us to the crucial need to consider how we might end the bifurcate tendencies that have historically driven relational divide based on race, ethnicity, sexuality, class or other modes of what we might consider difference. But if we look closely at these communities, the information they provide also signals a potential to produce joy out of deep duress and anxiety. This joy, what we call life, can help us recapture a sense of being that exceeds the languages of duress, be they technologically driven or otherwise.

Sheena Calvert: That's a beautiful way for us to conclude, because the idea that we need to start pre-empting our own futures rather than allowing those futures to just happen to us is a really important thought. Our conversation has really helped me to think about how we can pull back and think about our relation to technology (and everything around it), in a different way, that we don't have to just say, everything's doomed and it's all going to hell, but that we need to be cautious, we need to be thoughtful. We need to move towards the light.

Ramon Amaro: Exactly. We have to move towards the light of our own aspiration, even if these lamps seem dim and impossible to reignite in the present moment. So when I look at the data centre, I look through it as something that I have already seen, and at the same time my view is turned towards a correction of the paths already travelled.

OVH SBG2 data centre in Strasbourg destroyed by fire on 10 March 2021.
Photo by Abaca Press / Alamy Stock Photo

1. This text is exerpted from an interview that took place between Ramon Amaro and Sheena Calvert on 2 July 2023, in The Hague, the Netherlands.
2. See N. Schrader, 'Acid Clouds', in this book.
3. See brief video interview with R. Amaro, *Acid Clouds*, 13 May 2019. Online. Available at *www.acidclouds.org* (accessed 11 October 2023).
4. In other words, there is no 'view from nowhere' since I am part of the world and as such can achieve no objective distance from it. My search for objective truth is always compromised by my own place within the system I am seeking to achieve a distance from.
5. Which is to say that the world is infinitely complex and any representation of it (in data / diagrams), can only suggest that complexity, not repeat it. The 'sophistication and wonder' of such data exceeds any representational system. It remains beyond reach.
6. The argument that is sometimes made is that such a form of retrospection involves taking data from what's been deposited in the past and creating new things from it, in the present. We see this in the various AI image creation systems such as Midjourney, which scrapes existing data / images from the Internet and produces 'new' images from it. Everything is therefore produced from something that already exists. Hence: retrospectively.
7. Hyperstition is a term originating from theorists Nick Land and Mark Fisher. It involves a kind of self-fulfilling prophesy. By stating something in the form of a story, and by repeating those stories, this process brings about the reality. An often-cited example is the conspiracy theory, which takes on a life of its own through repetition and dissemination.
8. I. Kant, *Critique of the Power of Judgement*, P. Guyer (ed.) and P. Guyer and E. Matthews (trans.) (Cambridge: Cambridge University Press, 2000 [1790]).
9. See T.W. Adorno, *Negative Dialectics* (New York: Routledge, 2004 [1966]).
10. F. Fanon, *Black Skin, White Masks*, C. Lam Markmann (trans.) (London: Pluto Press, 2008 [1986]).
11. 'If negative dialectics demands the self-reflection of thinking, then this implies in tangible terms, that thinking must, nowadays at any rate, in order to be true, also think against itself. If it does not measure itself by the extremity, which flees from the concept, then it is cast in advance in the same mold as the musical accompaniment, with which the SS was wont to drown out the cries of their victims.' Adorno, *Negative Dialectics*, op. cit. (note 9), 369.
12. See D.J. Haraway, *Staying With the Trouble: Making Kin in the Chthulucene* (Durham, NC: Duke University Press, 2016).
13. See R. Amaro, *The Black Technical Object, Machine Learning and Aspiration of Black Being* (London: Sternberg Press, 2022).
14. Haraway, *Staying With the Trouble*, op. cit. (note 12).
15. Monad is a term from Leibniz. A monad is a basic element from which other material properties and objects are derived ('extended' into time and space). These basic or 'primary' substances (such as atoms, souls and God himself) are irreducible. They become extended into the world and create other things as they do so, but they are in themselves the most fundamental elements that make up those things. Each monad is individual, but becomes a part of more complicated collections of individual monads.
16. See G. Deleuze and F. Guattari, *A Thousand Plateaus: Capitalism and Schizophrenia* (London: Bloomsbury, 2013). 'Lines of flight' are detours and connections that don't conform to a preexisting system or structure. They emerge through the process of making tangential and / or lateral links between seemingly unconnected thoughts, ideas, things (creating another dimension or 'plane'). This process constitutes a form of 'deterritorialization' that creates multiple avenues rather than following an expected path. Such a process produces 'multiplicities' that connect to other multiplicities and so on and so forth, in an infinite (creative) process.
17. See M. Bowden, *AI: Its Nature and Future* (Oxford: Oxford University Press, 2016) and *Artificial Intelligence: A Very Short Introduction* (Oxford: Oxford University Press, 2018).
18. Edict, as in the order implicitly given to us within culture and society.

The Knots of the Metaverse: On Data Infrastructures and Indigenous Rights

Marina Otero Verzier

The MCHAP 0780 is the largest quipu on display at the Museo Chileno de Arte Precolombino in Santiago, Chile. Composed of 586 camelid fibre knotted cords, organized into 8 sections of 10 sets, each with up to 13 sub-levels of information holding 15,024 items of data, this quipu ('knot' in Quechua) dating from c.1500 is one of the systems for recording information used by the Incas. 1

State officials or *quipucamayocs* made and kept these three-dimensional devices to store 'census records and tribute accounts to genealogies, poems, songs and the deeds and exploits of prominent empire figures'. 2 Each of the quipus' strings is capable of reflecting millions of different combinations of colour, fibre type and structure, and knot direction to hold an enormous variety and quantity of knowledge. Together, these physical features and sequences construct a binary-coded system, whose mathematical organization anticipates the operations of a computer. 3

The quipus allowed for the transfer of vast information throughout the Inca empire. They coalesced the polity's economy and storage system, facilitating the coordination of crop harvesting, cotton and wool production, raw material availability, and the production of stock across geographies, climatic regions and astronomical cycles. Part of an extensive infrastructure, the quipus' knots were entangled with an advanced transportation system and a structure of *collcas*, or warehouses, where products and materials were stored and redistributed. 4

Throughout history, empires have relied on communication networks to circulate products, information and knowledge, to manage resources and their complex societal organizations. Often laid out on top of each other, these apparatuses connect analogous territories enabled by distinct technologies and political ideologies. The Inca Trail is not an exception, and its remnants across the Chilean territory have been occupied and repurposed by colonial, modern and neoliberal infrastructures. Paths transitioned by *chasquis* – 'runners' or messengers who carried quipus – and llamas moving materials and products turned into an intricate global communication system dependent on the exploitation, management and supply of natural resources. To the Inca empire, the quipu was what data infrastructure is to contemporary global markets.

In this age when data centres encode and process immense amounts of information, quipus remain largely indecipherable. If their long-kept stories were to be thoroughly read, the quipus would transform the understanding of the Inca empire and its large-scale social organization, whose history (due to lack of written records) has been told primarily by Spanish colonizers. As with today's digital networks, disentangling the reach of the quipus' cords and their mesmerizing function involves decoding their knots, a task that remains before and comparable to comprehending the convoluted workings of contemporary digital infrastructures.

Yet, the intricacy of digital infrastructure is not the only challenge. The climate catastrophe makes it necessary to reinvent how data is produced, circulates and stored if the industry is to deliver the promises of the metaverse without causing planetary meltdown. Major players are investing in new data centre models. In this context, quipus are images of an alternative architectural paradigm for storing data – a sophisticated medium not dependent on vast electricity and land consumption, CO_2 emissions and mining. Quipus, however, also embody the political and social system of their times. They materialize indigenous resistance against colonial domination – a history that resonates across the many communities currently fighting for their sovereignty and survival against epistemic violence and extractivism carried out in the name of green, connected, datafied futures.

Extraction

Not far from the MCHAP 0780 quipu, an hour drive from Santiago to Valparaíso, a new system that links fibre-optic cables, data centres, electromagnetic waves and sites for resource mining is about to make the Andean region a communication hub and, inevitably, an epicentre of the struggles against extractivist economies. The Humboldt Cable, the first submarine cable between Latin America and

Knotted-cords accounting instrument, camelid fibre Inca (Arica), MCHAP 0780. ***www.chileantesdechile.cl****/en/vitrinas/chile-bajo-el-imperio-de-los-inkas/los-nudos-del-inka*

Oceania connecting Valparaíso and Sydney, will turn Chile into a preferred data centre location.[5] The country is also the second-largest producer of lithium, a critical element for phones, electric vehicles and data centre batteries, present in only a few places on Earth in high concentration. The *salares* in the Atacama Desert in Chile are one of them.

Ancestral cosmologies and the prospects for the planet's immediate futures collapse in the Salar de Atacama. The territory, once traversed by the Inca Trail, is occupied by one of the largest brine-based lithium mines in the planet. Site of human, animal, vegetal and microbial coexistence for thousands of years, the Salar is endangered by the lithium extraction processes involving the evaporation of millions of litres of water in the world's driest desert. Processes aimed at producing the lithium-ion batteries that power the uninterruptible supply of information on our screens – and the data centres where this information is stored.[6] Whereas mineral extraction has been an ongoing activity in the Atacama Desert – it is what attracted the Incas to the territory and instigated the domination of the existing Atacameño communities – the scale of today's operations is putting the entire ecosystem at risk of disappearing. Our connected and low latency futures are granted at the expense of the *salares'* environmental destruction and dispossession of the indigenous communities who have been their stewards for generations.

Atacama, despite its hyper-biodiversity, is instrumentally portrayed as a lifeless space. Disregarding indigenous cosmic vision and stewardship practices, mining companies treat the landscape as one of the biggest repositories of resources and, therefore, a major economic asset. Managers and engineers impose their logistical models on the land, following cravings for productivity and corporate profit. Extending over the *salares* and defining a map of mining concessions, the Cartesian grid spatializes their ambitions.[7] The grid creates a blueprint for the ground's exploitation. It also dominates the aerial perspective over the rectangular, colourful brine ponds whose beautiful shots smoothly circulate in media; and whose aesthetic experience eclipses the social and environmental destruction that tails them.

Regardless of the grids, the Atacameños know that the Atacama's ground is alive. They know that the minerals exported from the Atacama to other territories around the world are not lifeless matter.[8] They are not commodities – a designation that has been used to justify their extraction – but micro-organisms native to the desert and an intrinsic part of a living ecosystem. Indigenous leaders, such as Rolando Humire, are also aware of the rapid degradation of the *salares*, their inhabitants (bacteria, flamingos, humans, plants) and the symbiotic relations between them. Drought, contamination and death resulting from the extraction of lithium and other minerals in the region are a byproduct of the mining industry, the global dependence on energy and data, and the predictions that assume the inevitability of its growth.

Legal Battles

It would be naive to consider, however, that the landscapes of resource extraction are circumscribed to a Cartesian grid laid over the desert. Atacama's ground is as entangled as a quipu's knots with technologies we use, cars we drive, architectures we inhabit, data we use, public officials we elect. There are plenty of examples of these enmeshments. Early in 2022, Chile's Ministry of Mining awarded two lithium extraction contracts to China-based BYD Chile SpA and Chile-based Servicios y Operaciones Mineras del Norte S.A. to extract up to 80,000 metric tonnes of metallic lithium (80 million kilograms). Organized through an international call for proposals, the tender sought, according to president Sebastián Piñera's right-wing government, 'to increase lithium production in Chile, thus meeting the growing global demand generated by the development of areas such as electromobility'.[9] Various societal sectors opposed the action carried out controversially in the last months of Piñera's government, right before the new administration led by left-wing politician Gabriel Boric took power on 11 March 2022.

Yet the lithium tender process had been long planned by those with interests in the business. Years before the tender resolution, BYD electric buses, a gift from China to Piñera's government, started to populate Chileans' streets.[10] China's leading manufacturers of lithium batteries and battery-powered electric vehicles, BYD is the company for which BYD Chile SpA is a subsidiary and which was awarded one of the lithium extraction contracts. Controlling lithium's extraction will mean controlling the entire market, from mine to end product.

The introduction to evermore sophisticated electric cars and autonomous transportation in cities, in turn, will demand data centres that are able to support the constant, real-time transfers required to relay information to users via smartphone apps powered by lithium batteries. BYD (also known as 'build your dreams') sells products that generate a demand for other BYD products, and its automated factories run at the rhythm of always anticipated growth.[11] That rhythm connects the smart city's promises with the natural resource extraction that is destroying ecosystems such as the Atacama, a reminder of what it takes to 'build our dreams'.

Today's data infrastructures are intertwined across geographies mines, fibre optic cables, switch-points, mobile telephone towers, automated ports and factories,

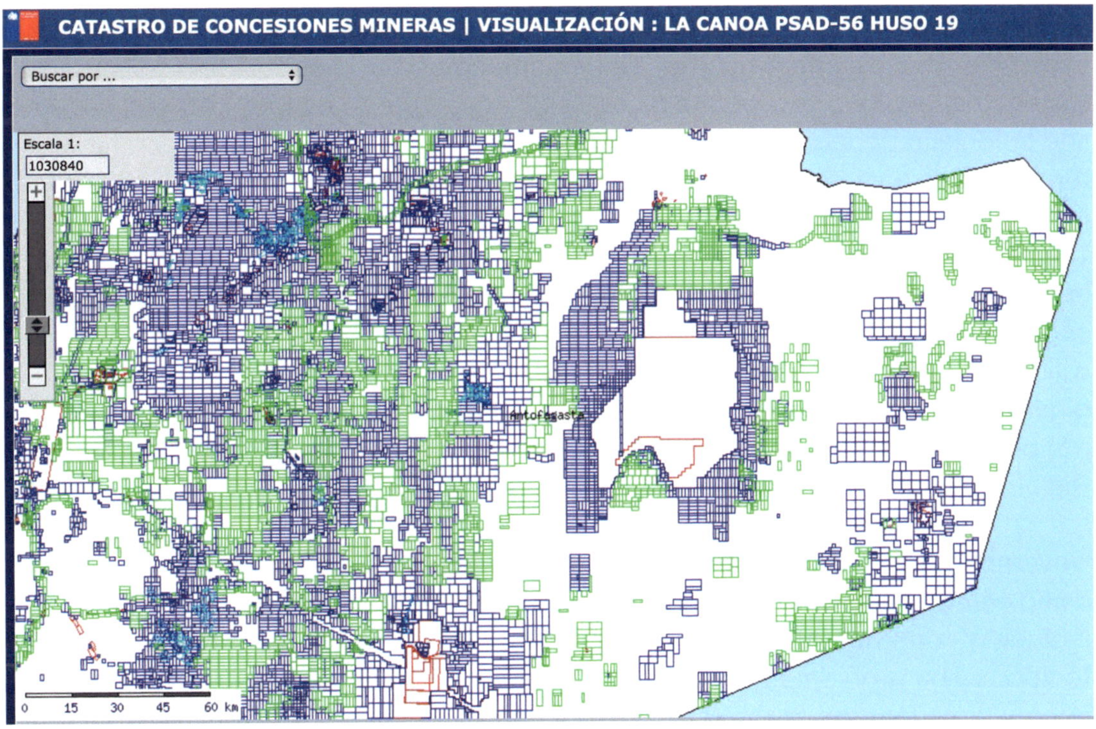

Catastro de Concesiones Mineras, Chile, mining concessions in the salar de Atacama, screenshot.

and, by extension, the space of everyday life. 12 Contemporary data-driven society embodies what I call the 'Cartesian enclosure', a spatial and epistemic system that extends beyond the more visible grid. One shaped by patriarchal and colonial rationales, their knots coalesce in seemingly unassuming architectures such as data centres. 13 Despite their apparent banality, these architectures are critical for current and future political, cultural and socio-economic activities. Their functioning depends on ever-growing extractive practices and increasing land and energy consumption. Developments in AI, the Internet of Things and the metaverse that demand vast amounts of data stored for extended periods will only make percentages rise. 14 We are contributing to an unprecedented digital-data production aimed at creating alternative, future worlds that are not only outpacing the scalability of today's storage solutions; they are putting the only world we all inhabit at risk.

Governments, pressured by local communities, environmental movements and rising energy prices have started to impose controls. Several data centre hubs including countries such as Singapore and the Netherlands have implemented temporary bans on data centre construction due to their excessive land and energy consumption at the detriment of nearby residents. 15 On 1 June 2022, Chile also took measures affecting the lithium supply chain and consequently the extraction of a main component for data centre batteries. Following a series of appeals filed by the Atacameño indigenous communities of Camar and Coyo alleging the violation of the right to live in a pollution-free environment, the Third Chamber of the Chilean Supreme Court annulled the bidding process for lithium exploitation. 16

With its decision, the Supreme Court confirmed that any intervention on ancestral territories or that may affect indigenous communities requires consultation in accordance with the Convention 169 in Article 5 of current Chile's constitution. 'It is a historic fact, unprecedented from the judicial point of view and from the perspective of the defence of the ancestral rights of the Atacameño communities that the Supreme Court accepts these appeals, defends these communities, gives value to their ancestral and territorial rights and in the defence of the Salar de Atacama and, therefore, suspends the contracts, the bidding and the awarding', declared Gabriel Muñoz, the lawyer representing the Atacameño Community of Coyo. 17 The day after, the government announced the constitution of a national lithium company and a lithium institute aimed at holding debate regarding this strategic element – a proposal already included in Boric's electoral program – particularly stressing the importance of dialogue with the indigenous peoples and the governance of the salt flats where lithium is found. 18

'Do I drink lithium? Do I eat copper?',
San Pedro de Atacama, Chile, 2022.
Photo by Rodolfo Muñoz

Predictions Matter

These and other episodes stimulate the demand for new paradigms for procuring, consuming and storing energy and data. They stress the need for renewed relations between digital infrastructures and indigenous peoples. Forecasts, however, continue to say otherwise. 19 According to IDC's Global DataSphere, data generated globally is expected to grow at 23 per cent annually from 2020 to 2025, reaching 180 zettabytes in 2025. 20 Gartner, in its report 'Market Trends: Evolving Enterprise Data Requirements', refers to a coming Zone of Potential Insufficiency and estimates that hyperscale vendor data storage demand could potentially soar to 50 per cent from 2020 through 2030. 21 Growth, according to these estimates, seems inevitable. And growth incentivizes markets. Instead of conceptualizing changing data and energy usage patterns, companies and governments keep on looking for 'less harmful' innovations attuned to land availability and the growing energy price. As a result, data centre architecture becomes a critical site of investment.

It is not a coincidence that the largest commercial real estate services company globally, CBRE, recently launched a data centre division. 22 Architecture and engineering firms such as Corgan, HDR, Gensler, AECOM and Arup are also expanding their data centre divisions. Underwater and seaborn data centres; hologram, fluorescent dye, crystal and DNA data storage systems; macro data centres ceding terrain to decentralized edge computing and micro centres; and quantum are only some of the proposals making the news. None of them, however, emphasize the impact of data centres, digital infrastructure and renewable energy projects on indigenous rights and sovereignty. None account for more-than-human relations, and data production and consumption behaviours that reduce dependence on growth. As the case in Chile has shown, those are crucial pursuits, particularly in a moment of expanding production and consumption of data and shaped by future predictions fuelling the extraction of more resources, the placement of new submarine cables, the construction of more data centres, the production of more batteries, often to the detriment of indigenous communities across the world.

Future Storage

Atacama is not the only site where the development of the digital infrastructure and 'green energy futures' comes at the expense of indigenous peoples. In the Nordic countries, for instance, the expansion of wind farms is having a dramatic effect on the environment and culture of the Sámi. The booming data centre industry in the region has been enabled by favourable tax conditions and the expansion of wind farms and other renewable energy initiatives. The presence of the wind turbines, however, is harming reindeer herders. By invading their pastures and displacing the animals from their usual grazing lands, the wind farms are endangering the protected cultural practice of traditional reindeer herding, essential to Sámi identity. 23 As with lithium mines, wind farms are part of the decarbonization efforts essential to tackle the worst impacts of climate change. However, in what many have described as 'green colonialism', these projects often ignore the rights of indigenous communities and endanger their survival in the name of sustainability.

Some judges still make a difference. In October 2021, after a series of rejected appeals in local courts, the Norwegian Supreme Court ruled that a series of turbines part of the largest onshore windfarm in Europe, violated Article 27 of the UN international covenant on civil and political rights. According to the article, minority ethnic people 'shall not be denied the right, in community with the other members of their group, to enjoy their own culture, to profess and practise'. 24 The Norwegian and the Chilean ruling demonstrate the importance of consultation and collaboration with indigenous and local communities in ensuring climate justice. Steps that governments and companies must yet usually do not follow.

As a result, indigenous communities do not aways wait for the other's goodwill and decide, instead, to develop their own programs. Stewart Stacey, managing director of Binary Security, made the news by starting the world's first indigenous-operated data centre at Charles Darwin University in Australia's Northern Territory capital, Darwin. 25 This initiative instigates indigenous governance over the creation, collection, ownership, stewardship, analysis and application of data. It adds to the work of collectives such as the Maiam nayri Wingara Aboriginal and Torres Strait Islander Data Sovereignty Collective working in support of indigenous data sovereignty and data governance protocols. 26 Their efforts are aimed towards exercising control of the data ecosystem and infrastructure to empower sustainable self-determination and effective self-governance of indigenous peoples and First Nations.

Binary Security will not be alone in its pursuit. Formed in 2018, Kalinda IT Services Pty Ltd, an indigenously owned Australian IT services business has recently partnered with Trifalga DC to develop a network of hyperscale and edge data centres across Australia, of which Toowoomba (a Queensland-based Pulse Data Centre) is their first location. 27 'We have a clear vision to drive digital inclusion, on country, in country and with country, and to create education and employment opportunities for Indigenous people within the IT, IoT and cybersecurity sector', Kalinda IT founder and chief executive Michael Dickerson explained. 28

Photo by Ivan Alvarado/Reuters/FastMedia

165

These cases demonstrate the strategic role of indigenous and subaltern communities in developing their own infrastructures and how data centres are relevant for the future of data sovereignty. Designers like Ibiye Camp also contribute to this shift. As a cultural, economic and strategic asset, data could be, according to Camp, generally mobilized to the detriment of marginalized communities. Yet it could also be, she claims, a vehicle for restitution for citizens during this time of overseas technological investments.[29] In her work, Camp explores how 'data materializes in the West African landscape and analyzes the spatial consequences of its production, consumption and storage'.[30] The African continent has a shortage of data centres, partially due to power outages and the reliance on polluting and expensive diesel generators. As a result, most of Africa's data is stored in Marseille.[31] Still, Nigeria, a country with the larger number of data centres in West Africa, is attaining significant investment in new infrastructures and data centres from overseas organizations. These developments, Camp shows, disregard the realities of the city's inhabitants, and lack ethical processes around data collection and management. Users are obliged to register biometric data in order to use digital infrastructures and telecommunication, raising concerns about governmental and corporate surveillance and control of Western companies over data, one of the legacies of colonialism.

In this environment, characterized by the conflicts between government, private corporations and citizens, data ownership is a fundamental asset. Camp's design strategy aims at intervening in this landscape through data collection devices that can be used by local populations in public spaces in Lagos.[32] Her project instigates the creation, collection, ownership and application of data as a cultural, strategic and economic asset in the hands of the population instead of corporations. She argues for the incompleteness of data and the unintentional disruptions in digital technology as a form of resistance to counteract the multinationals' control over the city and its inhabitants. In that sense, Camp's work resonates with other initiatives that enact the decentralized trust-based networks historically used by African communities and take advantage of the advancements in edge computing and micro data centres to create autonomous systems. Not dependent on big corporations and able to be installed almost anywhere, these facilities are popular across libertarians, remote communities and activist groups.

These and other episodes remind us of the longstanding battlers sustained by indigenous and subaltern communities throughout history. In these struggles for their sovereignty, their lives and rights, the control of information and suppression of knowledge has been a consistent strategy of the oppressor; the violence is often directed – as we have seen in the case of the by Spanish conquistadors against the Incas – towards the communities' communication means.

The MCHAP 0780 quipu, ascertained to be around 450 years old, was found in an Incan cemetery in Arica, Chile. 33 A meticulous study of its knots, arranged in decimal order, allowed for the tabulating of 15,024 units in different categories whose meanings are still unknown. 34 In the early moments of the colonization, Hispanic writing was prioritized over the quipus' system of knotted strings, which was largely illegible to the colonizers. 35 Later, in around the 1570s, the quipus would achieve colonial recognition as records and would start to be admitted in legal and administrative contexts. 36 Andean litigants could then 'read' the strings in lawsuits concerning mistreatment by the colonial society such as tribute restitution. 37 Granting legitimacy, however, was part of the policy adopted by Francisco de Toledo, the new viceroy, who sought the establishment of order in the Andes to optimize the extraction of resources on a national scale. 38 These included forced relocations of Andean populations to Hispanic-type settlements and the establishment of the system of *cabildos* (municipal councils) where scribes would transcribe the information of each community's quipus. 39

During that time of intercultural communication, quipus would also provide strategies for daily resistance against colonial rule, subverting the Catholic imposition of confession, or facilitating the planning of uprisings through quipu missives, illegible to colonial officials. 40 Similar episodes triggered what has been described as the colonizer's 'fear for the knotted rope', which ultimately resulted in the prohibition of the quipu by the Catholic Church on 23 December 1583, during the Third Session of the Limense Provincial Council. 41 Not many survived. And despite the occasional transcription of accounting quipus by the colonial administration, these records were not kept together with the quipus or their illustrations, thus rendering the relations between lasting material evidence and its meaning unreadable.

Paradoxically, today's digital infrastructure is seen as a key to unlock the quipus' secrets and protect them from decay and the passing of time. Developments in AI and other digital tools are, to many scholars, crucial in the efforts to decipher quipus and their combinatorial iterations of colours and knots. The more than 850 digitalized quipus around the world are a de facto database that could be studied resorting to trained computer algorithms and macro analysis. 42 Large-scale patterns and similarities could emerge from the study, across national borders, of these digital and reproductions by anthropologists, archaeologists, historians, linguists, mathematicians.

And yet the attraction to the enigmatic presence of quipus and the desire to unpack those chronicles of the Andean past embodied in their knots resorts to the same logics that put at risk the very landscapes and communities where quipus were born. Academic and scientific efforts in cataloguing, digitizing, analysing and ultimately deciphering quipus, are part of a historical lineage that connects with the European colonial order and its material, cultural and epistemological genocide in favour of the hegemony of Occidental knowledge. 43 Transforming the quipus in digital, Cartesian databases implies subjecting, once more, indigenous knowledge and cosmovision to the Western knowledge-power apparatus. Their knots link those bodies, ecosystems and knowledges erased by colonization with those at risk of disappearance by contemporary extractivist violence.

Against the compulsion to extract the quipus' information and render it transparent, the acceptance of their opacity and incommensurability stands as a decolonial practice – a form of resistance against the totalizing Western European gaze imposed on others. 44 Only then could we recognize that what was destroyed by colonization was not only the knowledge to code and decode quipus, but the situated social, cultural relations organized around them, leaving the surviving knotted cords as empty signifiers and evidence of the violence of erasure unleashed by colonialism.

Reconsidering the relevance of a data-driven society and its expansion at the expense of indigenous communities becomes paramount. Only by doing so could we undo the knots that tie up the mining industry and extractivist economies with logistical and communication networks and human dreams with the logic of growth. And perhaps, then, embrace a different way of being in the world.

Not far from the lithium brine pools where mining companies extract almost 2,000 litres per second from the deepest layers of the desert and where thousands of millions of litres of water are evaporated annually to run and host the metaverse, the Atacameño communities subsist through ancestral solidarity. 45 Many of the *ayllus* (family clans) and villages of San Pedro de Atacama date back 12,000 years, making it an area of historical importance. 46 Traditional architecture was conceived to be attuned to all the beings in *ayllu* and orientated towards Licancabur, the volcano, who marks the start of the day and the start of life, revealing the sunrise every morning. Before construction, Atacameños make an offering to the land, ask permission and express gratitude. By doing so, the communities confront the notion that someone can actually 'own' a place, its resources and export them in exchange for profit. 'We don't own anything. We are part of everything', explains Rolando Humire. It is only with permission from the spirits, the ancestors and Mother Earth, *Patta Hoiri*, that the Atacameño community interacts with the land; their activities are not only based on 'taking' but primarily on caring and 'giving' back. 47

As of land, the relation with water is essential for the life in Atacama and manifests in non-Cartesian eco-social systems. In villages such as Toconao, waters are communal, and customary indigenous norms apply for their administration. Irrigation is organized in turns. 48 Water is treasured, cared for and redistributed to all members of the community through a series of water canals and ponds, whose maintenance is organized by the *comunidad de regantes*. To prevent damaging effects on the ecosystem, no one uses more water or during a different turn. Limitless growth there and anywhere else has never been an option. Notwithstanding the historical importance of the Chilean Supreme Court ruling against the lithium bidding process, in Toconao the struggle continues: as the metaverse and its low-latency digital subjectivities expand elsewhere, the communities' irrigation turns are inevitably set further and further apart.

Toconao, San Pedro de Atacama, Chile, 2022. Photo by Marina Otero Verzier

1. 'Inca Knots', Museo Chileno de Arte Precolombino, website. Online. Available at *www.chileantesdechile.cl/en/vitrinas/chile-bajo-el-imperio-de-los-inkas/los-nudos-del-inka* (accessed 2 February 2022).
2. 'Record Keeping With Knots In *Quipu*', *Counting With Knots in the Inka Empire*, exhibition website, Museo Chileno de Arte Precolombino, Santiago. Online. Available at *www.precolombino.cl/wp/en/exposiciones/exposiciones-temporales/exposicion-quipu-contar-anudando-en-el-imperio-inka-2003* (accessed 2 February 2022).
3. Whereas the theories comparing the registration system in quipus with binary code had an enormous influence in recent decades, it also faced criticism by authors such as Galen Brokaw. See Manuel Medrano, *Quipus: Mil años de historia anudada en los Andes y su futuro digital* (Lima: Planeta Perú, Kindle Edition, 2021), 25, 32, 135.
4. In Quechua, *collca* or *Qullqa* also names the Inca deity Qullqa, patron of warehousing and preserving seeds.
5. 'Desarrollo País elige a H2 Cable como socio estratégico para desarrollar Cable Humboldt', in Subsecretaría de Comunicaciones de Chile, 17 December 2021. Online. Available at *wwww.subtel.gob.cl/desarrollo-pais-elige-a-h2-cable-como-socio-estrategico-para-desarrollar-cable-humboldt* (accessed 17 December 2021).
6. For an account of the implications of lithium extraction at the Salar de Atacama see F. Díaz, A. Kubrak and M. Otero Verzier (ed.), *Lithium: States of Exhaustion* (Rotterdam/Santiago: Het Nieuwe Instituut/Ediciones ARQ, 2021).
7. See mining concessions in the Salar de Atacama. Catastro de Concesiones Mineras, Chile. Online. Available at *catastro.sernageomin.cl* (accessed 28 September 2022).
8. The work of Marisol de la Cadena is relevant in this context. De la Cadena's theories on indigenous political strategies question modernity. Her work invites exceeding Cartesian categories to embrace the ways of being 'between' and 'with'. According to this cosmovision, mountains are not lifeless sources of materials, but earth-beings. See M. de la Cadena, *Earth Beings: Ecologies of Practice across Andean Worlds* (Durham, NC: Duke University Press, 2015).
9. 'Mining Ministry Notifies the Awarding of a Public Tender to Increase Lithium Production', in Gob.cl, digital portal of the Government of Chile, 12 January 2022. Online. Available at *www.gob.cl/en/news/mining-ministry-notifies-awarding-public-tender-increase-lithium-production* (accessed 15 January 2022).
10. 'BYD renueva su compromiso para impulsar la electromovilidad en Chile', BYD, 14 January 2020. Online. Available at *www.bydchile.com/info-noticias.php?not=63* (accessed 10 February 2022).
11. Elsewhere I have written about BYD-automated factories. See M. Otero Verzier, 'Nothing Is Automatic: Producing More-Than-Human Relations in the Pearl River Delta', *AD Special Issue: Production Urbanism: The Meta-Industrial City* (September/October 2021) vol. 91, no. 5, 32–39.
12. Analysing the material implications of data infrastructures was one of the aims of the architecture studio ADS8: Data Matter: Digital Networks, Data Centres & Posthuman Institutions, led by K. Hilmi Dalkir, M. Otero Verzier and I. Pestellini Laparelli at the Royal College of Art, London, 2018–2020. Online. Available at *www.rca.ac.uk/study/schools/school-of-architecture/architecture/ads-themes-2019-20/ads8-data-matter-digital-networks-data-centres-posthuman-institutions* (accessed 28 September 2022).
13. See M. Otero Verzier, 'Cartesian Enclosures', in: M. Mahdavi and L. Wang (ed.), *New Geographies, 12: Commons* (Cambridge, MA: Harvard University Press, 2022), 39–57.
14. G. Kamiya et al., 'Data Centres and Data Transmission Network, Tracking Report – November 2021', International Energy Agency, website. Online. Available at *www.iea.org/reports/data-centres-and-data-transmission-networks* (accessed 18 January 2022).
15. Recently, plans by Facebook parent company Meta to construct the Netherlands's largest data centre in the town of Zeewolde, a small community of 20,000 people, faced opposition by residents and the Dutch senate. Given the growing opposition, Meta decided to pause the project. The controversy prompted the Dutch government to announce a nine-month ban on new hyperscale data centres until stricter regulations and licensing processes are in place. 'Our space is limited, so we have to make the right choices,' said Minister of Housing and Spatial Planning Hugo de Jonge in a letter to the House of Representatives. See M. Meaker, 'Facebook's Data Centre Plans Rile Residents in the Netherlands', *Wired*, 7 January 2022. Online. Available at *www.wired.com/story/facebook-dutch-data-centre* (accessed 16 January 2022); see also 'Kabinet besluit tot aanscherping regels hyperscale data centres', report from the Netherlands government, 16 February 2022. Online. Available at *www.rijksoverheid.nl/actueel/nieuws/2022/02/16/kabinet-besluit-tot-aanscherping-regels-hyperscale-datacentres* (accessed 17 February 2022).

16. 'Corte de Apelaciones de Copiapó dicta orden de no innovar y suspende proceso de licitación del litio', Poder Judicial de Chile, 14 January 2022. Online. Available at www.pjud.cl/prensa-y-comunicaciones/noticias-del-poder-judicial/67845 (accessed 28 January 2022); see also 'Corte Suprema acoge recursos de protección de comunidades indígenas atacameñas por licitación del litio', *Diario Constitucional.cl*, 1 June 2022. Online. Available at www.diarioconstitucional.cl/2022/06/01/corte-suprema-acoge-recursos-de-proteccion-de-comunidades-indigenas-atacamenas-por-licitacion-del-litio (accessed 28 September 2022) and E. Carrizo, 'Corte Suprema acoge recursos de protección de comunidades indígenas atacameñas por licitación del litio', *latercera.com*, 2 June 2022. Online. Available at www.latercera.com/pulso/noticia/gobierno-da-fin-a-la-fallida-licitacion-del-litio-impulsada-por-la-administracion-del-expresidente-pinera/N2CCOKNY45EPHE22JI6AGQ5N74 (accessed 28 September 2022).
17. 'Corte Suprema falla en favor de comunidades indígenas y deja sin efecto bases de licitación del litio en Atacama', *El Monstrador*, 1 June 2022. Online. Available at www.elmostrador.cl/cultura/2022/06/01/corte-suprema-falla-en-favor-de-comunidades-indigenas-y-deja-sin-efecto-bases-de-licitacion-del-litio-en-atacama (accessed 28 September 2022).
18. E. Carrizo, 'Gobierno aclara que rol de privados en la industria del litio impulsada por el Estado será minoritaria', *La Tercera*, 6 June 2022. Online. Available at www.latercera.com/pulso/noticia/gobierno-aclara-que-el-rol-de-los-privados-en-la-industria-del-litio-impulsada-por-el-estado-sera-minoritaria/AMSJHTHUT5DFDA5PXGTFHKCIUI (accessed 28 September 2022).
19. Thomas Alsop, 'Number of Data Centres Worldwide in 2022, by Country', *Statista*, 26 January 2022. Online. Available at www.statista.com/statistics/1228433/data-centres-worldwide-by-country (accessed 17 February 2022).
20. J. Rydning and D. Reinsel, 'Worldwide Global StorageSphere Forecast, 2021–2025: To Save or Not to Save Data, That Is the Question', IDC, March 2021. Online. Available at www.idc.com/getdoc.jsp?containerId=US47509621 (accessed 28 September 2022).
21. 'Gartner Market Trends: Evolving Enterprise Data Requirements – How Much Is Not Enough?', 14 July 2020, Gartner. Online. Available at www.gartner.com/en/documents/3987548 (accessed 28 September 2022).
22. Data Centre Solutions, CBRE. Online. Available at www.cbre.com/services/property-types/data-centre (accessed 6 February 2022); see also Peter Judge, 'Cbre Launches Brokerage for Data Centre Investors', *Data Centre Dynamics*, 1 April 2019. Online. Available at www.datacentredynamics.com/en/news/cbre-launches-brokerage-for-data-centre-investors (accessed 16 December 2022).
23. Andrew Lee, 'Naive Wind Industry Could Destroy Our Way of Life', *Recharge News*, 20 February 2019. Online. Available at www.rechargenews.com/wind/naive-wind-industry-could-destroy-our-way-of-life/2-1-547258 (accessed 28 September 2022).
24. International Covenant on Civil and Political Rights, adopted on 16 December 1966 by the General Assembly resolution 2200A (XXI), United Nations Human Rights, Office of the High Commissioner. Online. Available at www.ohchr.org/en/instruments-mechanisms/instruments/international-covenant-civil-and-political-rights (accessed 28 September 2022); see also Agence France-Presse in Oslo, 'Norway Court Rules Two Windfarms Harming Sami Reindeer Herders', *The Guardian*, 11 October 2021. Online. Available at www.theguardian.com/world/2021/oct/11/norway-court-rules-two-windfarms-harming-sami-reindeer-herders-turbines-torn-down (accessed 28 September 2022).
25. 'From the Land to the Cloud ... And Beyond. Darwin Is Set to Become Home to the World's First Indigenous Operated Data Centre', press release, Binary Security, 21 December 2020. Online. Available at journalists.medianet.com.au/DisplayAttachment.aspx?j=942900&s=2&k=1337381 (accessed 2 February 2022).
26. See Maiam nayri Wingara, website. Online. Available at www.maiamnayriwingara.org (accessed 2 February 2022).
27. 'Trifalga Teams up With First Nations Data Centre Operator', *Australia Property Journal*, 13 December 2021. Online. Available at www.australianpropertyjournal.com.au/2021/12/13/trifalga-teams-up-with-first-nations-data-centre-operator (accessed 28 September 2022). Maiam nayri Wingara means 'to welcome good knowledge' in the languages of the Mer, Palawi Kani and Darug peoples. Maiam nayri Wingara Indigenous Data Governance protocols and principles, Mayi Kuwayu. Online. Available at www.mkstudy.com.au/indigenousdatasovereigntyprinciples (accessed 28 September 2022).

28. N. Arboleda, 'Kalinda It Picks Pulse Data Centres as First Partner for Data Centre Rollout', *CRN Australia*, 14 April 2022. Online. Available at *www.crn.com.au/news/kalinda-it-picks-pulse-data-centres-as-first-partner-for-data-centre-rollout-578753* (accessed 28 September 2022).
29. Ibiye Camp was one of the architects involved in the architecture studio ADS8 led by K. Hilmi Dalkir et al. cited above within which she initiated her project Data, the New Black Gold. Online. Available at *www.rca.ac.uk/study/schools/school-of-architecture/architecture/ads-themes-2019-20/ads8-data-matter-digital-networks-data-centres-posthuman-institutions* (accessed 28 September 2022); see also Ibiye Camp's contribution to *Dixit 02: A Matter of Data* (2021) edited by M. Otero Verzier and Y. Santos.
30. Ibiye Camp, 'Data, the New Black Gold', Royal College of Art. Online. Available at *www.rca.ac.uk/showcase/show-2019/schoolofarchitecture/architecture/ibiye-camp* (accessed 28 September 2022).
31. 'Data Centres Are Taking Root in Africa', *The Economist*, 4 December 2021. Online. Available at *www.economist.com/middle-east-and-africa/2021/12/04/data-centres-are-taking-root-in-africa* (accessed 20 December 2021).
32. Ibiye Camp, 'Data, the New Black Gold', *ibiyecamp cargo collective*. Online. Available at *www.cargocollective.com/ibiyecamp/Area-Snap-Devices-Data-The-New-Black-Gold* (accessed 28 September 2022).
33. Quipu of Arica, '*Quipu*: Counting With Knots in the Inka Empire – 2003'. Online. Available at *www.precolombino.cl/en/exposiciones/exposiciones-temporales/exposicion-quipu-contar-anudando-en-el-imperio-inka-2003/quipu-de-arica* (accessed 2 February 2022).
34. Ibid.
35. Medrano, *Quipus*, 6, 18.
36. Ibid., 71.
37. Ibid., 69.
38. Ibid., 71.
39. Ibid.
40. Ibid., 81, 84.
41. Ibid., 72, 87.
42. Ibid., 131.
43. Portuguese sociologist Boaventura de Sousa Santos coined the term 'epistemicide' in his book *Epistemologies of the South: Justice Against Epistemicide* (2014), to refer to the destruction of indigenous knowledge of the natural world during the Spanish conquest.
44. Consider the work of Édouard Glissant on the 'right to opacity' as an antonym of transparency. Glissant argued for accepting the unintelligibility that often characterizes cross-cultural communication, as opposed to the 'transparency' as an epistemological modality of colonial power.
45. In addition to salt water, mining companies use a considerable amount of fresh water for the processing activities and worker's maintenance. See Sophia Boddenberg, 'Chile: Explotación de litio deja sin agua a pobladores', *Deutsche Welle*, 27 January 2020. Online. Available at *p.dw.com/p/3WsYe* (accessed 2 February 2022).
46. The term *ayllu* is a traditional Andean social unit, often structured by family and kinship relations. See María Rostworowski, *Recursos naturales renovables y pesca: siglos XVI y XVII* (Lima: Instituto de Estudios Peruanos, 1981), 39–43.
47. See 'Introduction' and 'On the Ground. Rolando Humire, Cristina Dorador and Alonso Barros, in Conversation With Godofredo Pereira', *Lithium*, 9, 10, 31.
48. See I. Sepúlveda Rivera et al., 'Aguas, riego y cultivos: Cambios y permanencias en los ayllus de San Pedro de Atacama', *Estudios Atacameños* (2015) vol. 51, 185–206. Online. Available at *dx.doi.org/10.4067/S0718-10432015000200012* (accessed 28 September 2022).

Mapping Acid Clouds

In this section you will find a selection from the photo series *Acid Clouds* by Roel Backaert and Niels Schrader, including images of the making process and infographics on data centres in the Netherlands.

Bearing Witness: Acid Clouds Photography

Roel Backaert and Niels Schrader

The photo series *Acid Clouds* started when Niels Schrader was looking for visual material to illustrate a presentation on data pollution in the Netherlands to be given at the *Fault Lines Symposium* at the Royal Academy of Art in The Hague in 2018. The publicly available photos of endless server racks were not satisfying enough to give a truthful image of what the actual condition of the Dutch spatial data infrastructure looked like. Generic images of servers, network cables and cooling equipment organized in long corridors would not bring forward any operational details and any sense of physical geography that is hidden behind a veil of abstraction. It seemed like the vagueness of the images was serving a purpose, namely, to strengthen the myth of pristine data cathedrals and to divert public scrutiny away from what data centres really are: concrete capital-driven places of real estate, where corporations use large-scale computation to monetize user data.

This finding left Schrader and photographer Roel Backaert, who became involved in the project, with questions like: What do data centre facilities in the Netherlands really look like? Who operates them and who has, or doesn't have, access? What initially started as a one-off challenge, soon became an obsession. With endless determination, Backaert and Schrader searched the entire Netherlands to capture the physicality of these new infrastructures of money exchange, which not only manifest a new typology of architecture, but also foreshadowed an uncanny image of tomorrow's automated landscape.

Central to the project is Schrader's observation that in the information design discipline the amount of data has been increasing exponentially in the last decades. What originally would fit on a portable medium like a floppy disk or a CD would later become stored on seemingly intangible platforms like FTP servers or the cloud. In the process, users have lost the sense of the physicality of data and with it any oversight. What can be observed in individual practices within the creative sector, applies in fact to the society at large. New efficient digital production processes are increasingly prevalent, driven by the capitalistic promise of infinite economic growth.

During the nighttime photo shootings, teamwork was central. Not every security guard wants to see two strange characters sneaking around their data centre, balancing on a ladder that overtowers the security fence and looking for the perfect angle for a photo. So, to minimize the exposure and attention-grabbing behaviour during their nightly expeditions, Schrader mapped the local conditions of the data centre locations ahead of time and then Backaert was able to capture the researched premises in his distinctive artistic style of colourful twilights. The resulting photos catalogue the material traces of virtual data in the Netherlands and give evidence of the physical impact of its urban condition.

The publication in front of you presents a selection of 101 data centres photographed by Roel Backaert in their spatial contexts, sorted by zip code. Through its extreme exposure time, the night-time photographs not only capture matters invisible to the human eye but also erase human actors from the canvas. After an exposure of many seconds, minutes, or even hours, colours start changing and more and more details emerge from the shadows. Direct lighting from urban infrastructure or secondary light reflected from clouds and aerosols brightens even unlit locations. This skyglow effect gives every location a unique colour pattern, and also creates a certain mood of atmospherical silence. Natural light sources like moonlight or starlight and the lack of people, intensify the eery effect.

The *Acid Clouds* photographs are taken from similar points of view, in order to methodically capture the spatial presence of server farms and enable visual comparisons across different topologies. By deliberately showcasing the images in the form of a series, unique traits of the buildings emerge, making apparent the motives and effects hidden in today's global cloud infrastructure. Presenting the photographs as a series reminds the audience also of the plural meaning of the word 'data', which, strictly speaking, signifies a plural form and is technically unquantifiable.

Since 2018, Niels Schrader and Roel Backaert have been systematically documenting Dutch data centres throughout the Netherlands.

Data centres presented in this publication are highlighted in bold and accompanied by the page number.

Dutch data centre locations, 2018 – 2024.
Source: Acid Clouds research

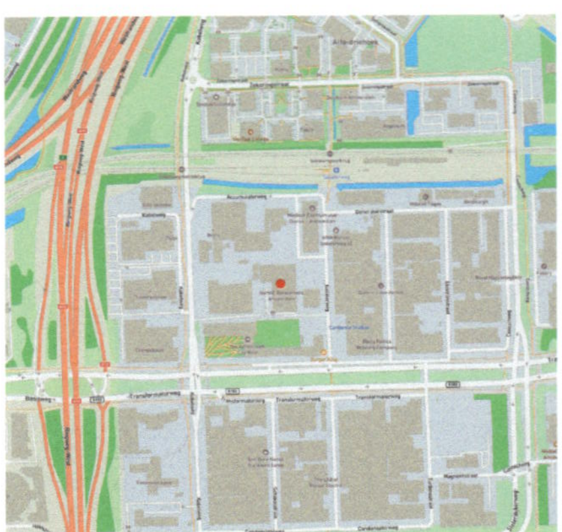

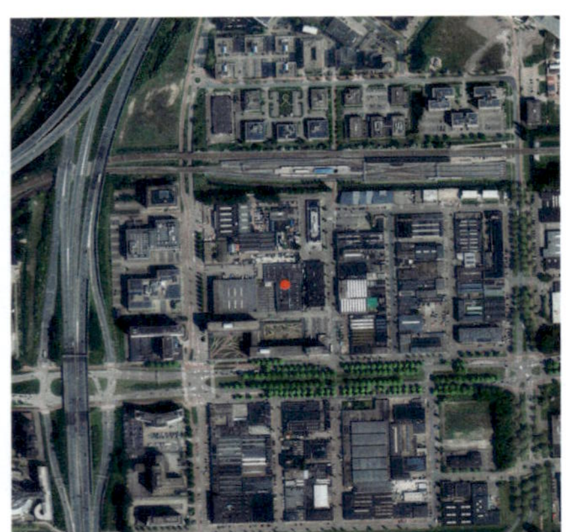

NorthC Datacenter Amsterdam
Kabelweg 48
1014 BB Amsterdam

Equinix AM8
Gyroscoopweg 2E
1042 AB Amsterdam

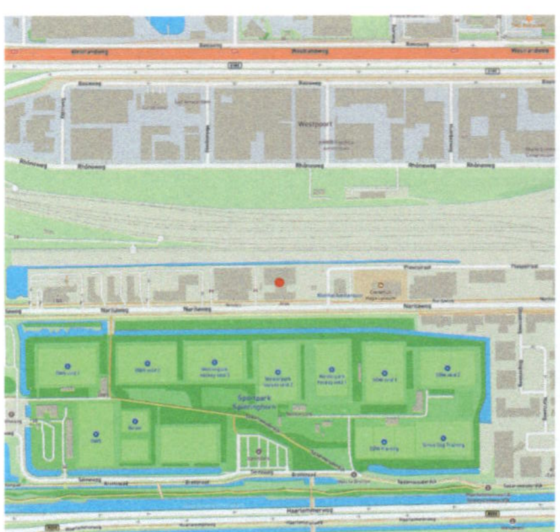

Digital Realty Data Center
Naritaweg 52
1043 BZ Amsterdam

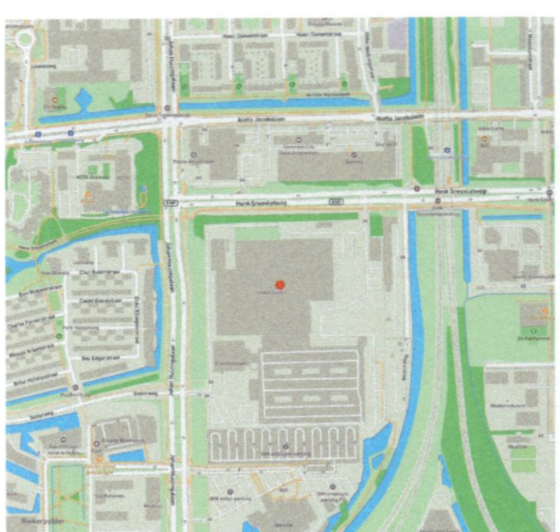

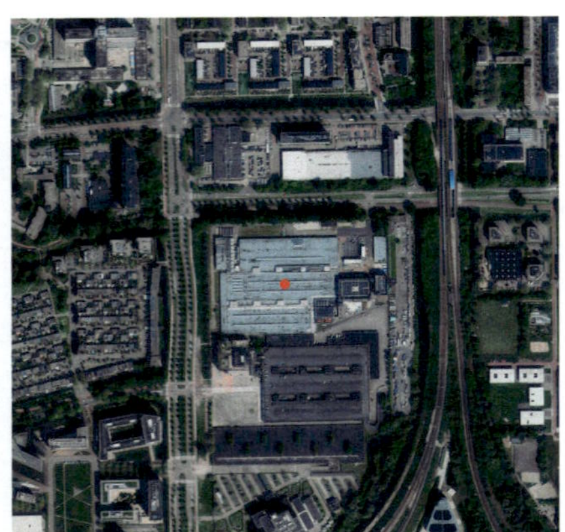

Global Switch
Johan Huizingalaan 759
1066 VH Amsterdam

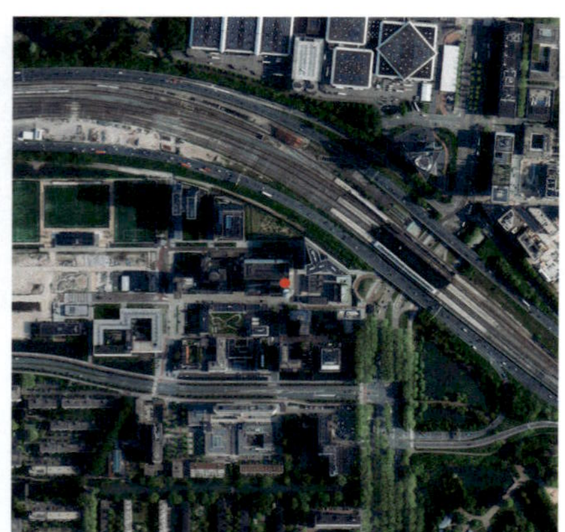

Cellnex Tower Amsterdam
Tommaso Albinonistraat 188
1083 HM Amsterdam

JCDecaux

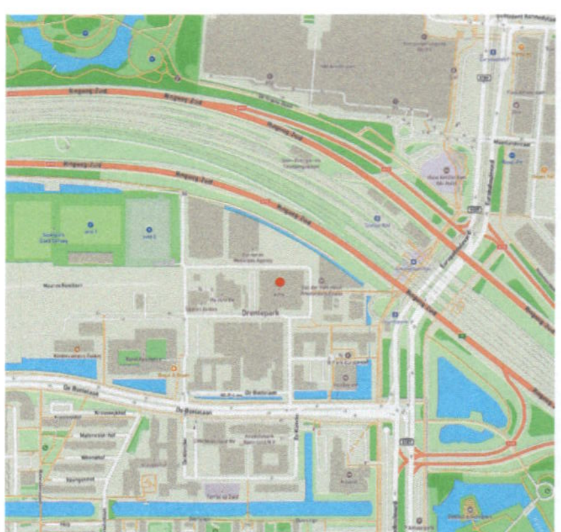

XS4ALL Datacenter DC 2
Barbara Strozzilaan 251
1083 HN Amsterdam

kpn

Equinix AM6
Duivendrechtsekade 80
1096 AH Amsterdam

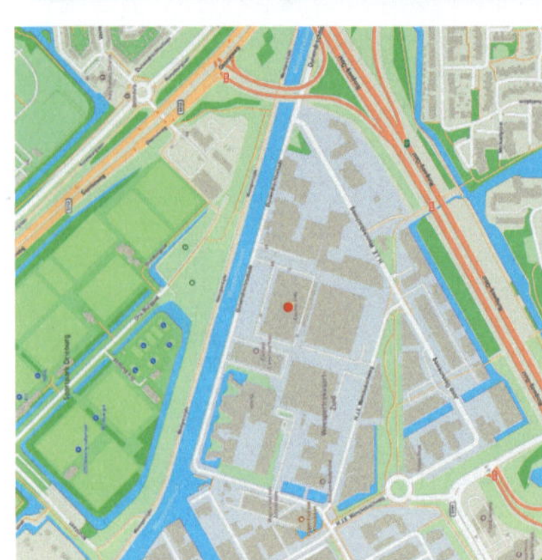

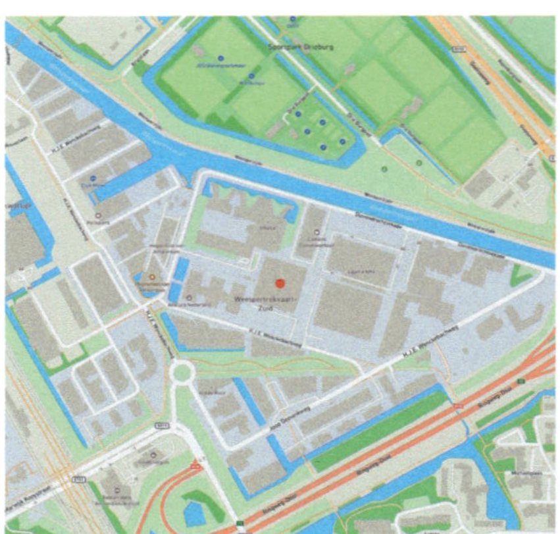

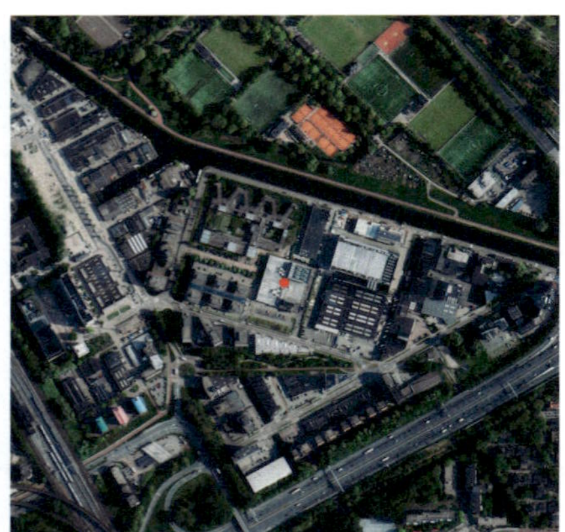

Digital Realty Data Center
H. J. E. Wenckebachweg 127
1096 AM Amsterdam

Digital Realty Data Center
Paul van Vlissingenstraat 16
1096 BK Amsterdam

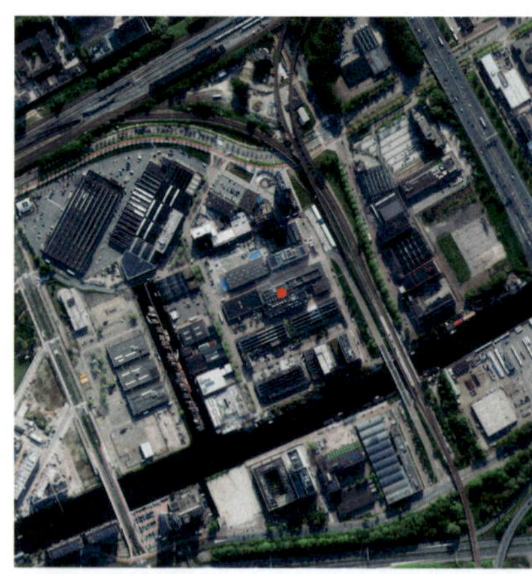

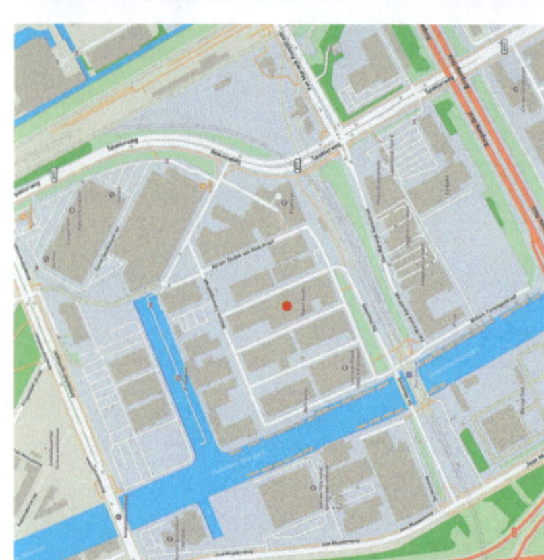

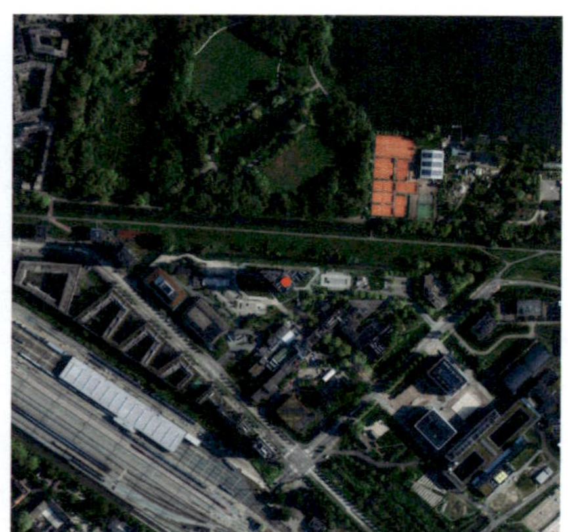

Digital Realty Data Center
Data Tower
Science Park 120
1098 XG Amsterdam

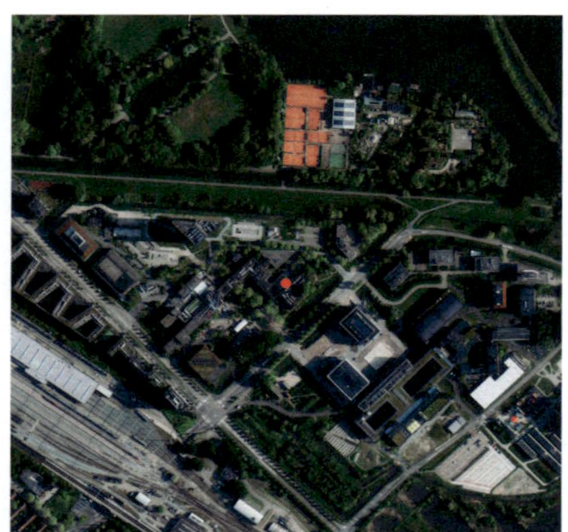

Interxion AMS9
Science Park 121
1098 XG Amsterdam

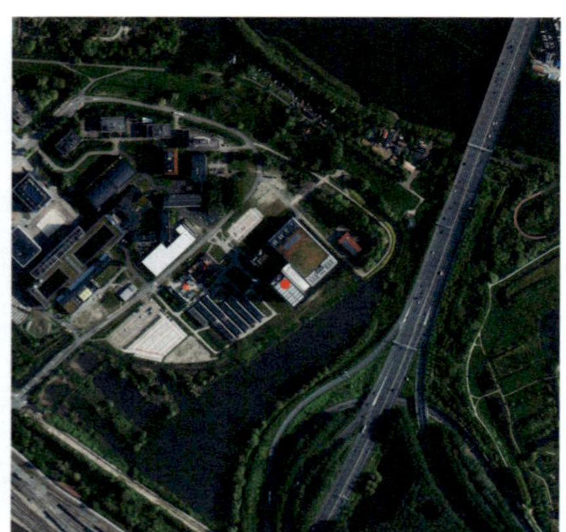

Equinix AM4
Science Park 610
1098 XH Amsterdam

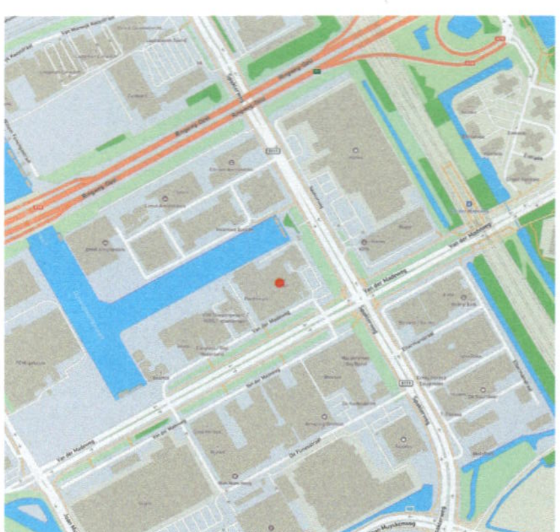

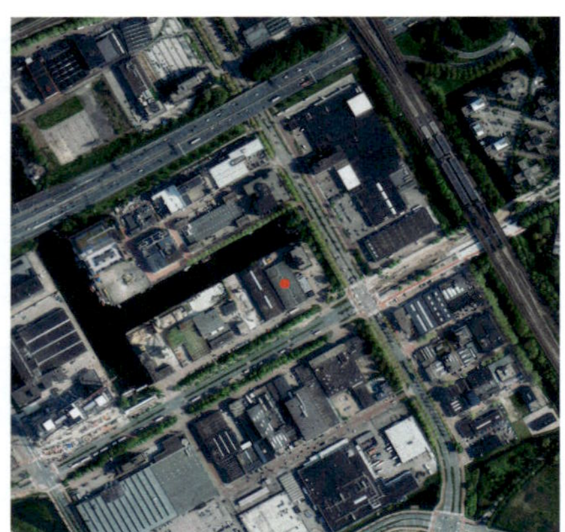

Colt Amsterdam Duivendrecht
Van der Madeweg 12–14a
1099 BT Amsterdam

colt

Equinix AM7
Kuiperbergweg 13
1101 AE Amsterdam

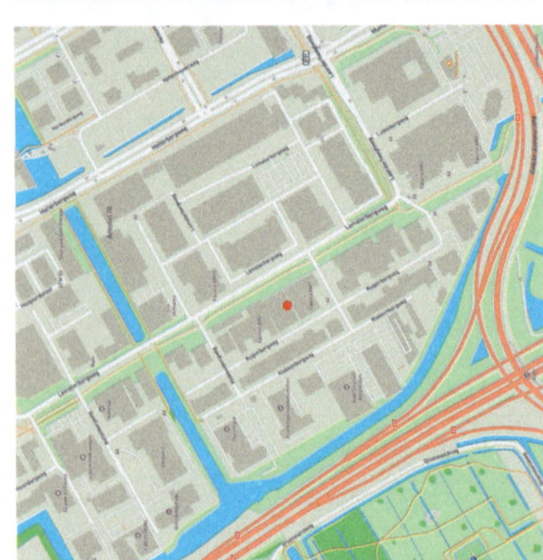

Equinix AM11
Lemelerbergweg 28
1101 AJ Amsterdam

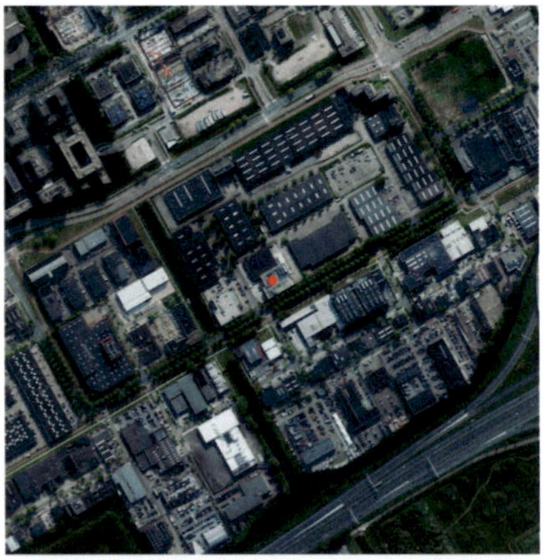

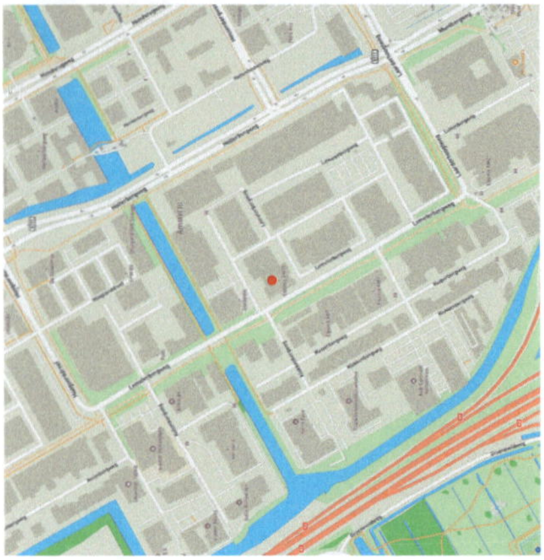

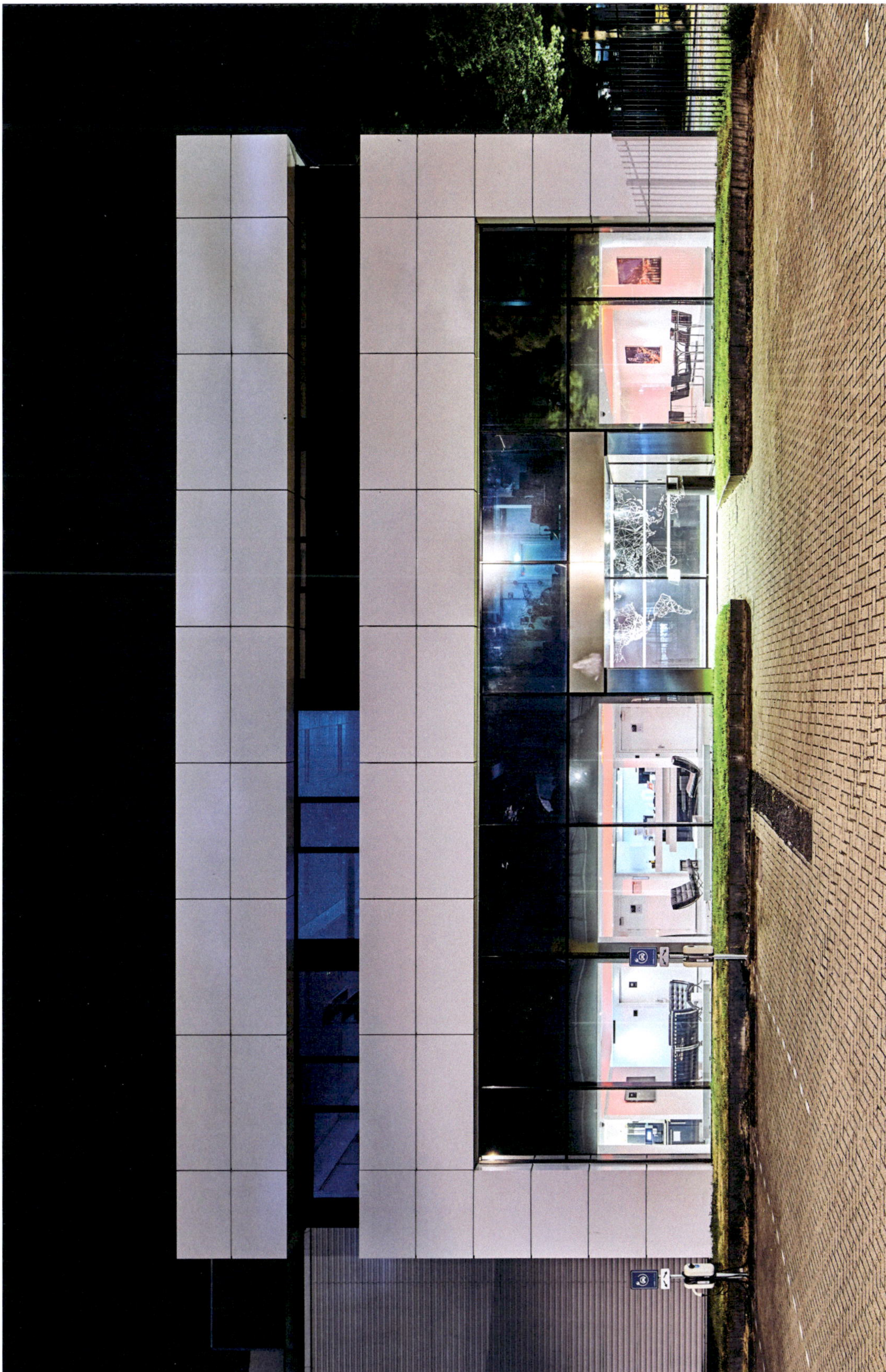

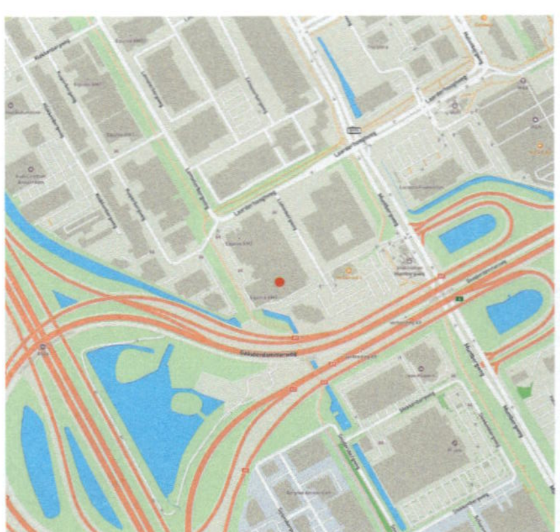

Equinix AM1
Luttenbergweg 4
1101 EC Amsterdam

Datacenter.com
Keienbergweg 22
1101 GB Amsterdam

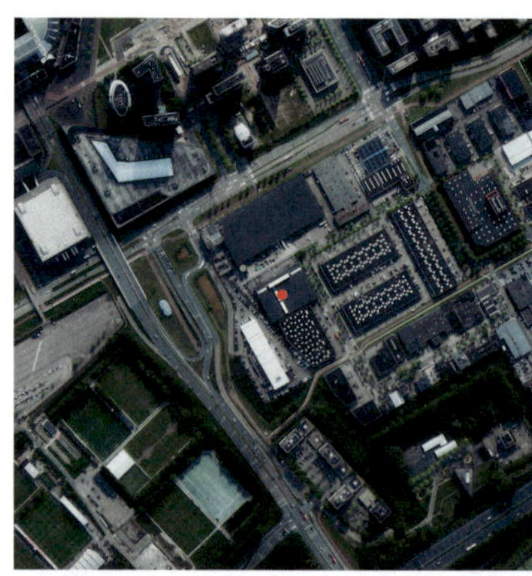

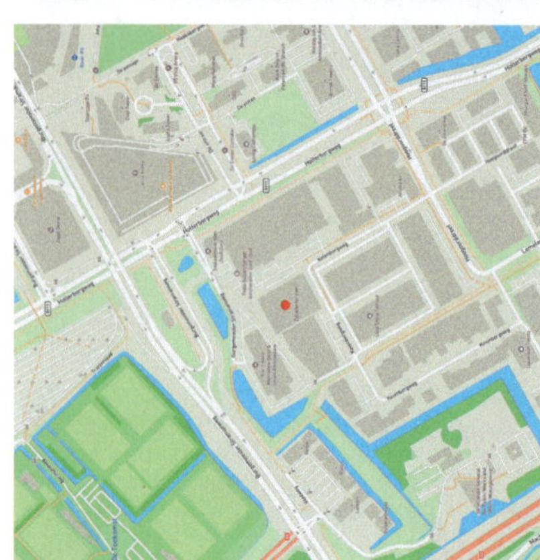

/// DATACENTER.COM

CenturyLink
Stekkenbergweg 4
1105 AJ Amsterdam

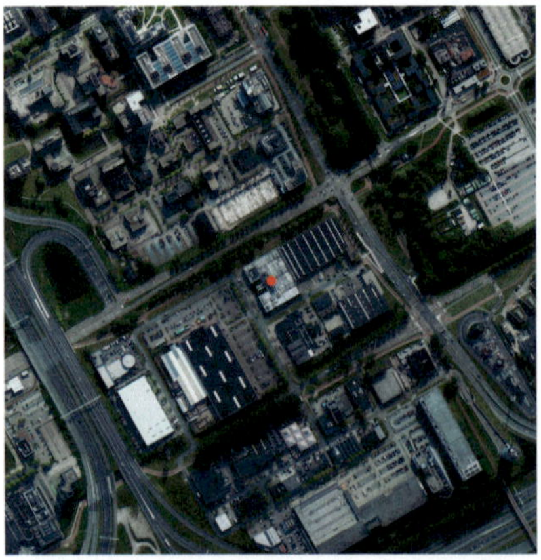

Equinix AM5
Schepenbergweg 42
1105 AT Amsterdam

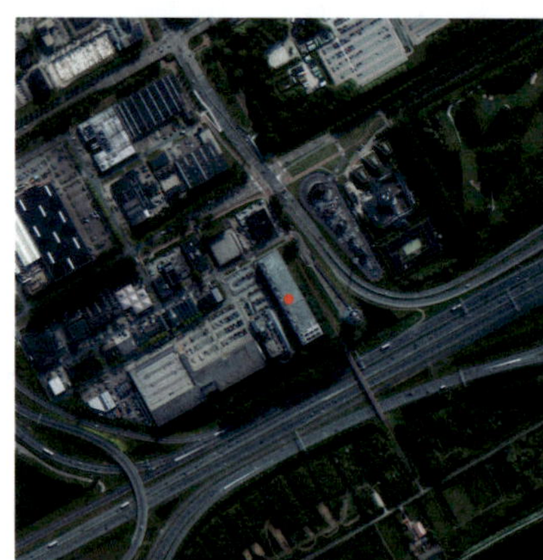

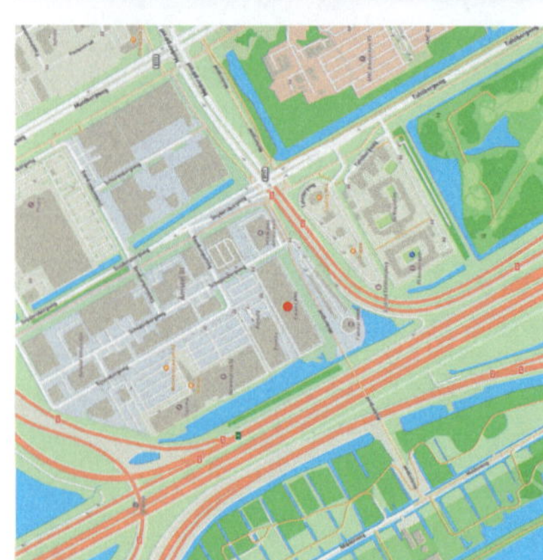

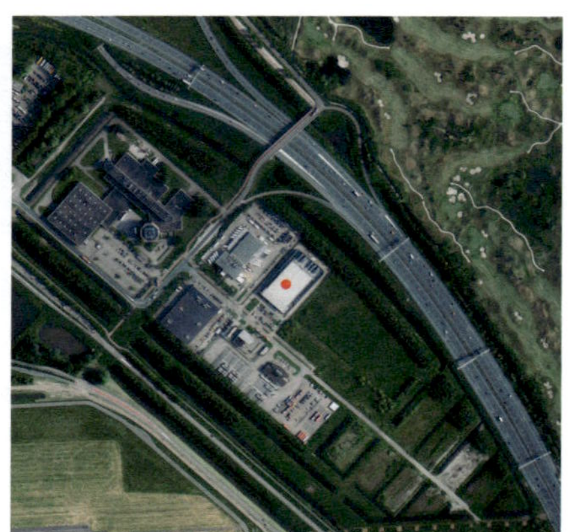

Digital Realty Data Center
NAP of Amsterdam
Cateringweg 5
1118 AM Schiphol

Interxion AMS6
Koolhovenlaan 25
1119 NB Schiphol-Rijk

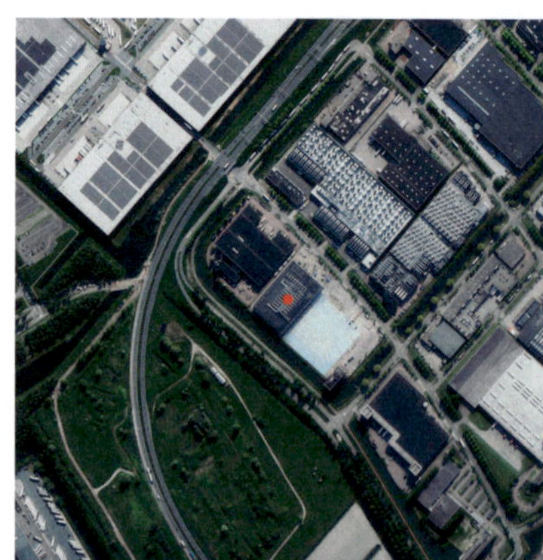

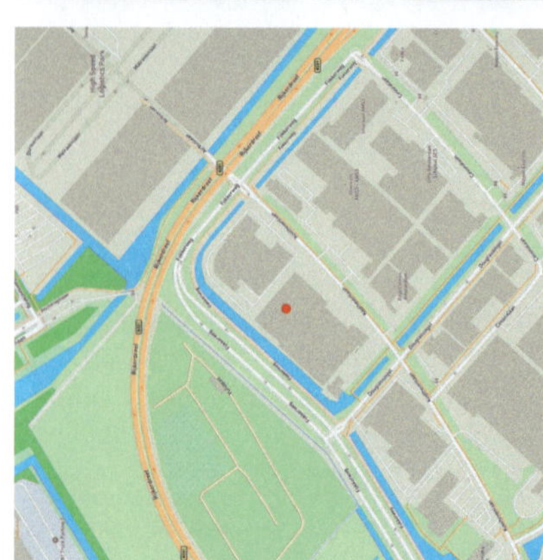

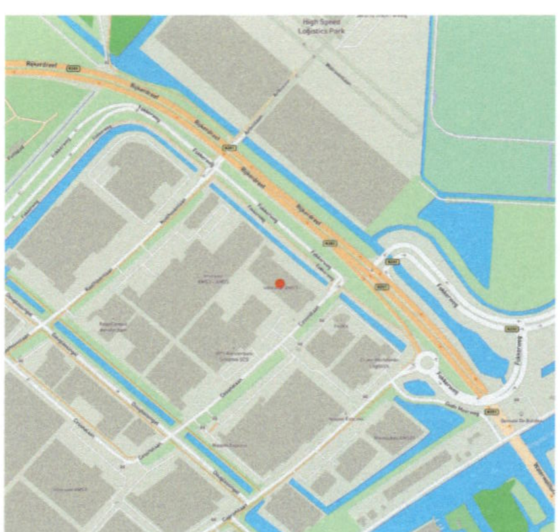

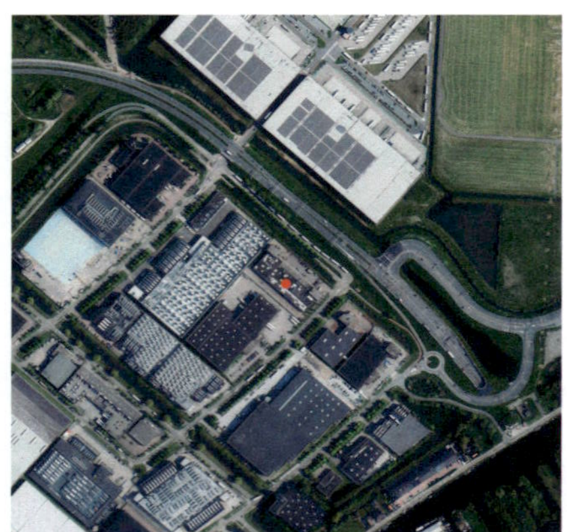

Interxion AMS3
Cessnalaan 1
1119 NJ Schiphol-Rijk

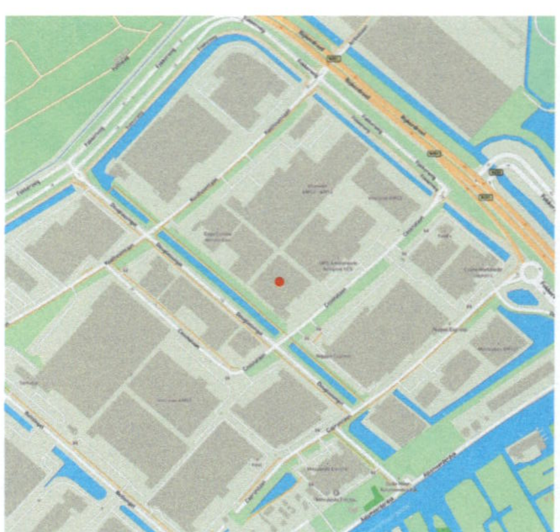

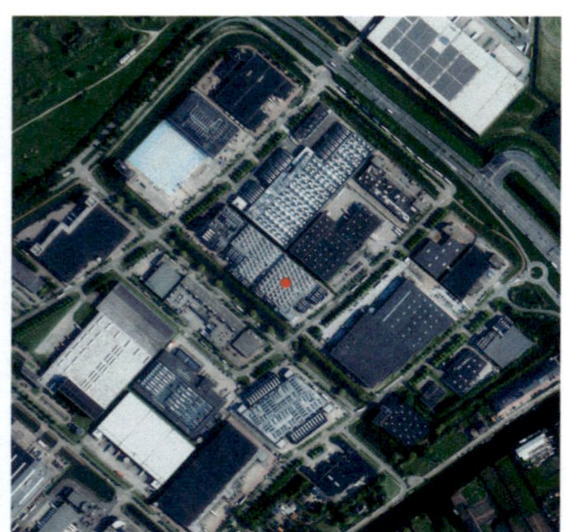

Edgeconnex AMS02
Cessnalaan 55A
1119 NK Schiphol-Rijk

Edgeconnex AMS04
Cessnalaan 32
1119 NL Schiphol-Rijk

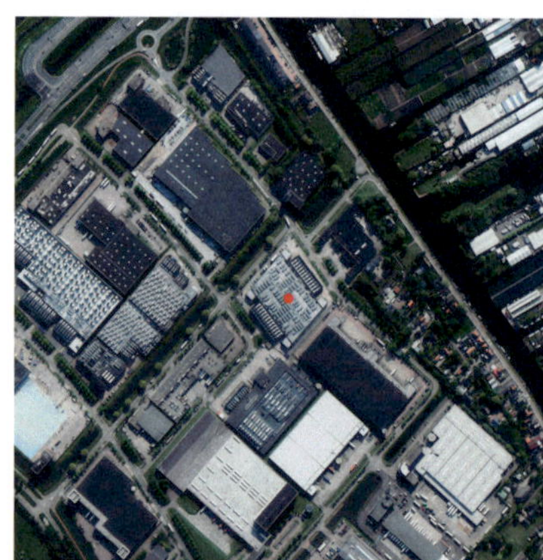

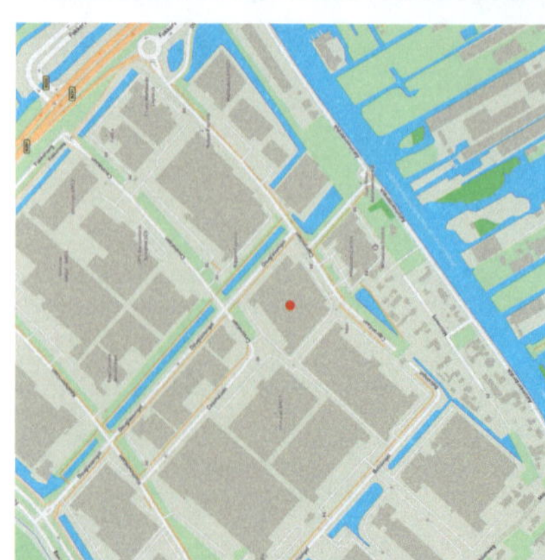

Interxion AMS7
Cessnalaan 50
1119 NL Schiphol-Rijk

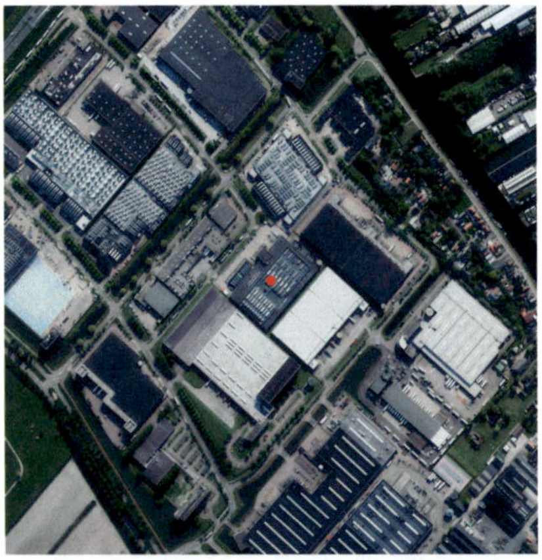

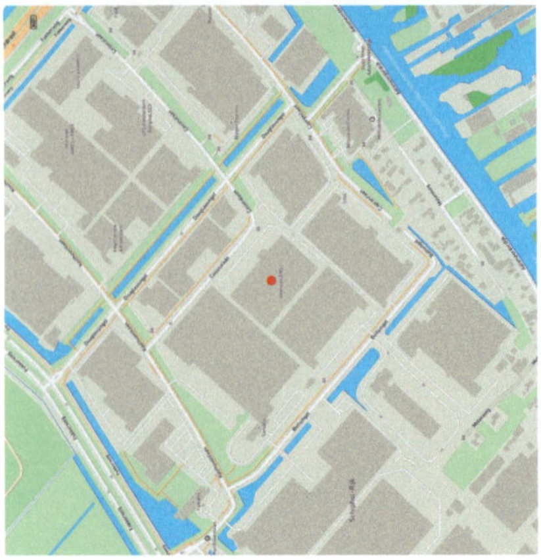

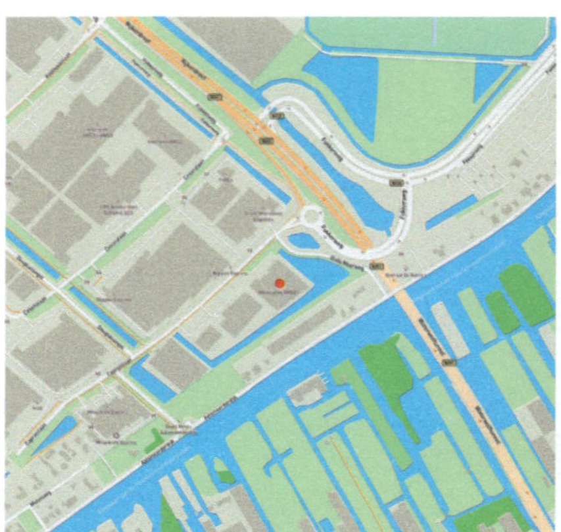

maincubes Datacenter
Capronilaan 2
1119 NR Schiphol-Rijk

maincubes

Interxion AMS5
Tupolevlaan 101
1119 PA Schiphol-Rijk

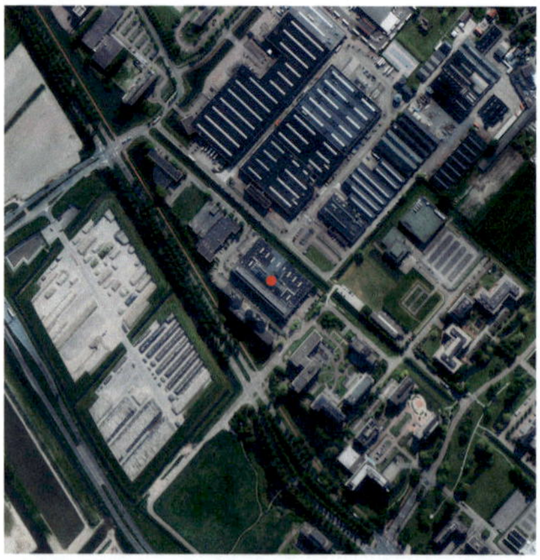

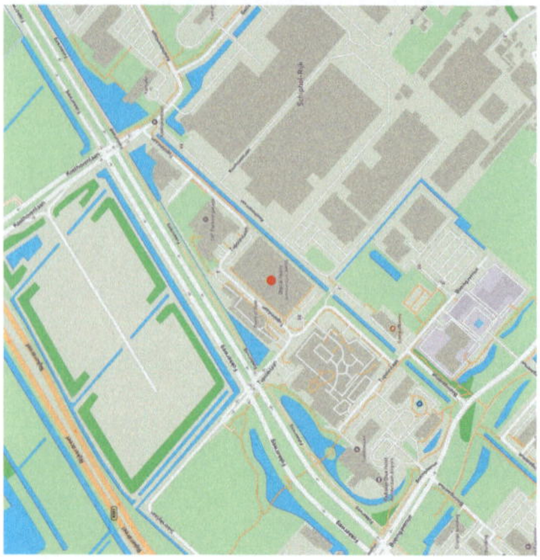

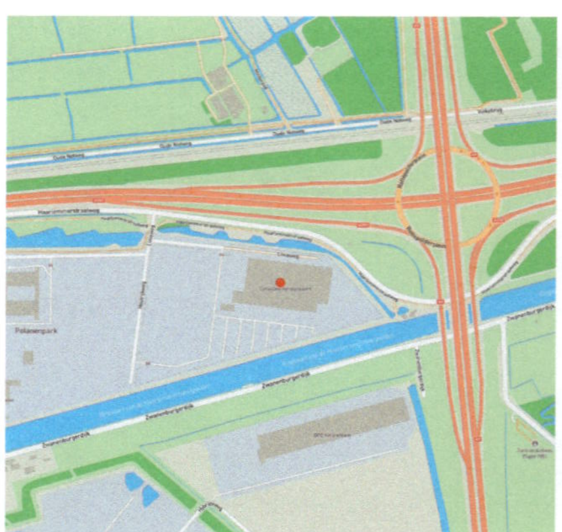

Cyrus ONE
Linieweg 1
1165 AA Halfweg

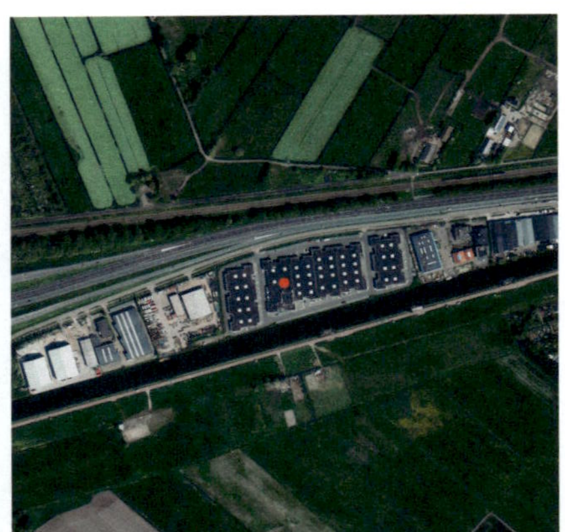

Cogent Communications Amsterdam
Haarlemmerstraatweg 135
1165 MK Halfweg

B11

Cellnex Tower Hilversum
Witte Kruislaan 47A
1217 AM Hilversum

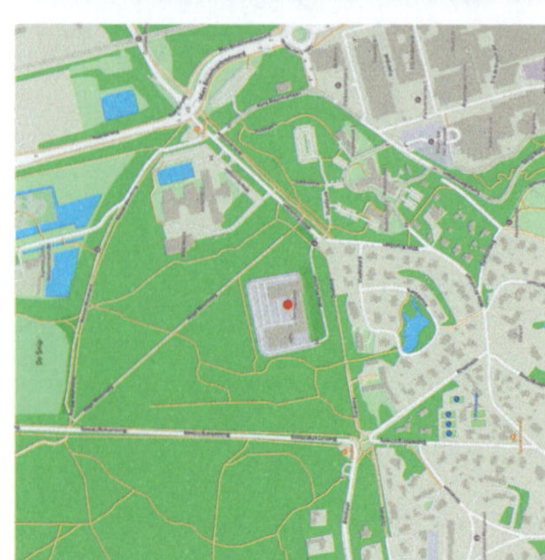

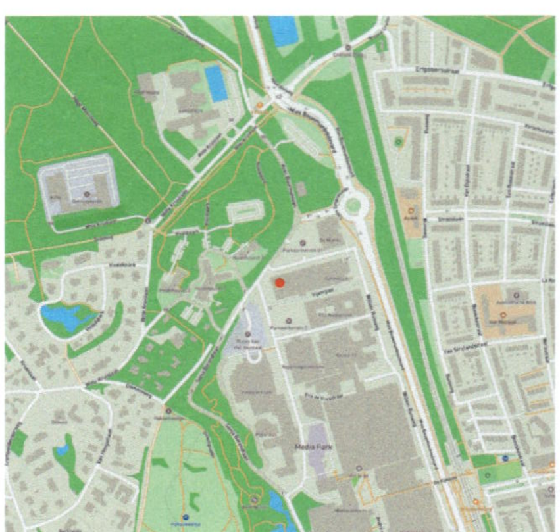

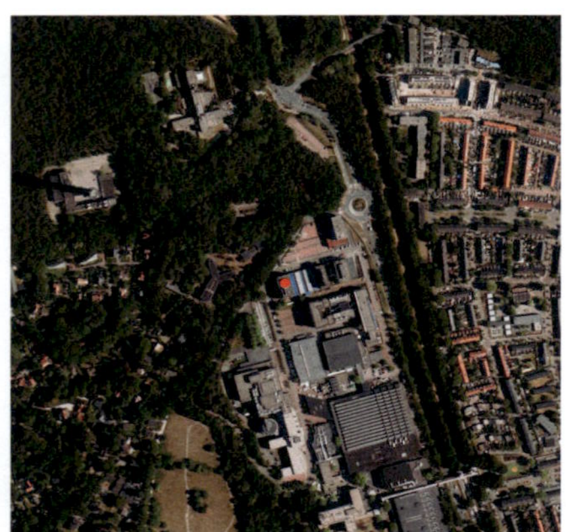

Red Bee Media
Koos Postemalaan 2
Media Park – Building Gateway C
1217 ZC Hilversum

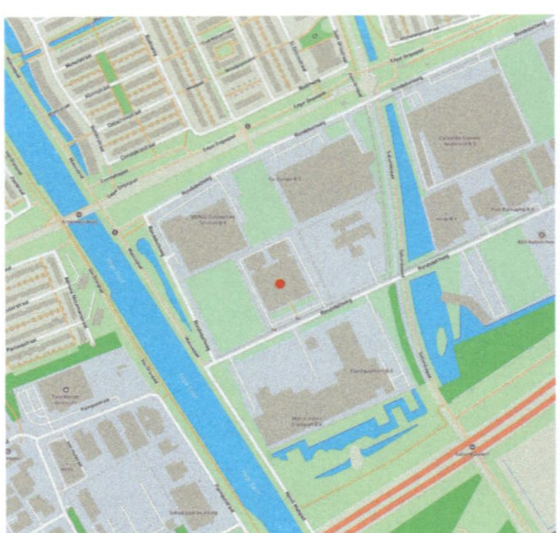

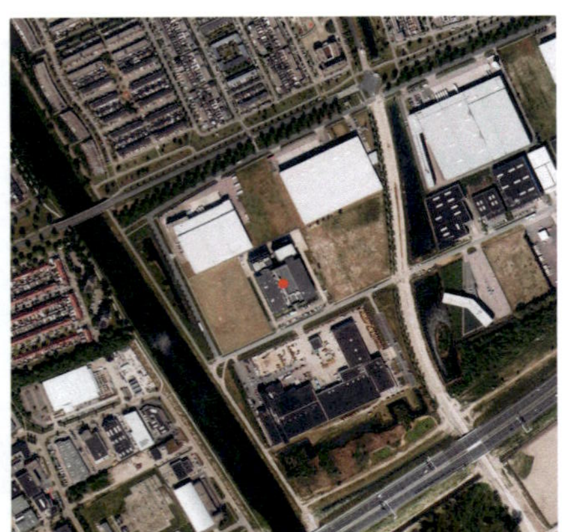

NorthC Datacenter Almere
Rondebeltweg 62
1329 BG Almere

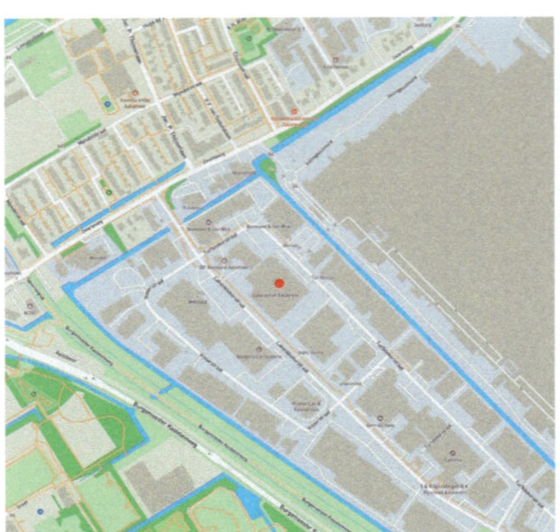

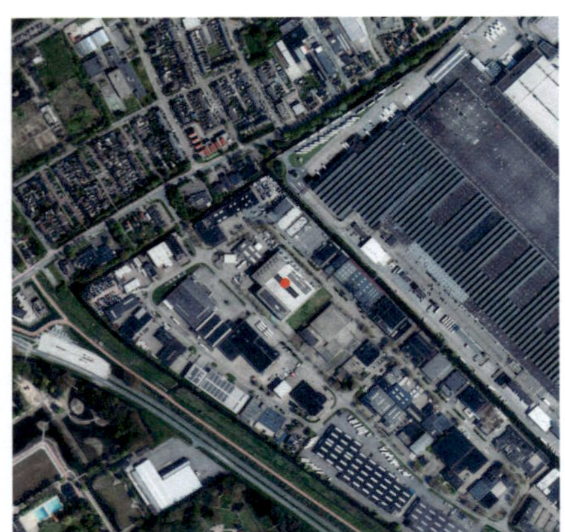

NorthC Datacenter Aalsmeer
Lakenblekerstraat 13
1431 GE Aalsmeer

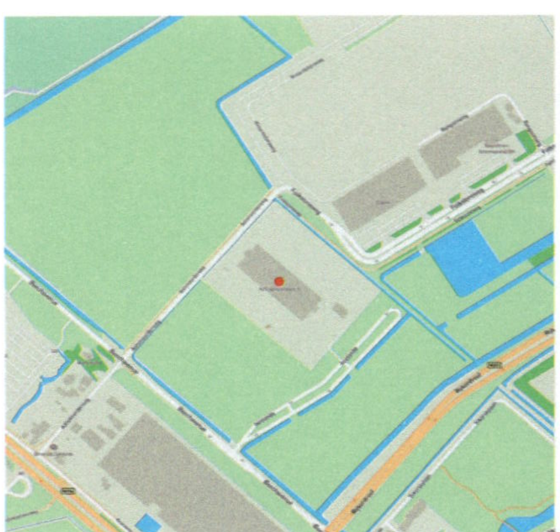

e-shelter Data Center
Aviolanda 1
1437 ED Rozenburg

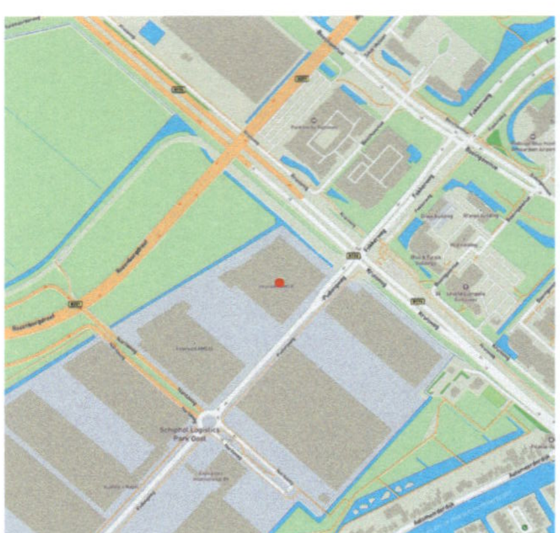

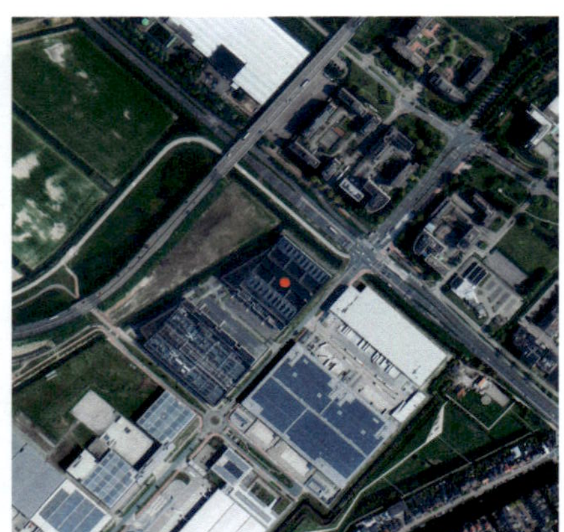

Interxion AMS8
Pudongweg 37
1437 EM Rozenburg

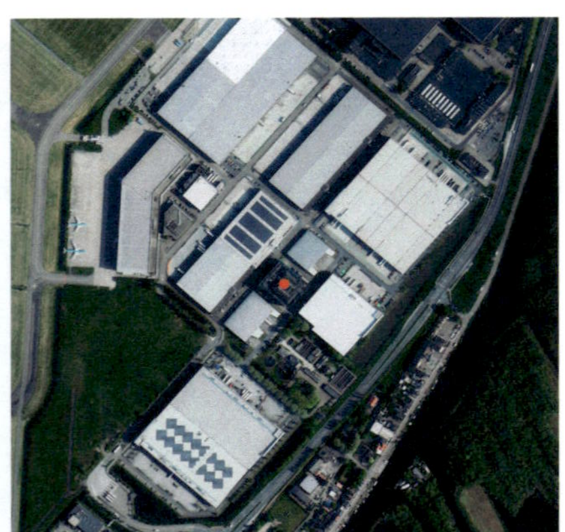

NorthC Datacenter Oude Meer
Fokkerweg 300
1438 AN Oude Meer

CTS GROUP
aramex
aramex

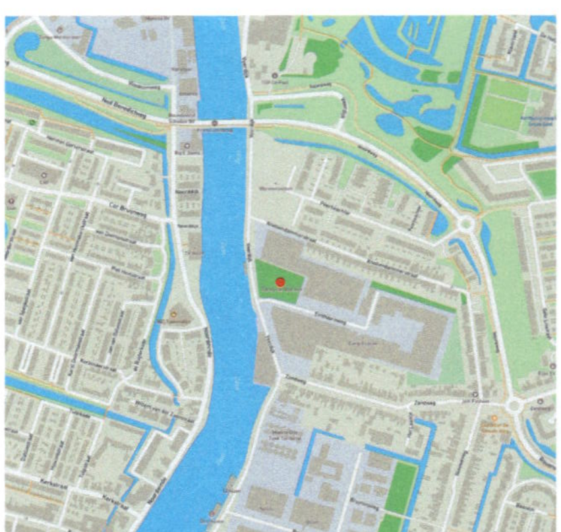

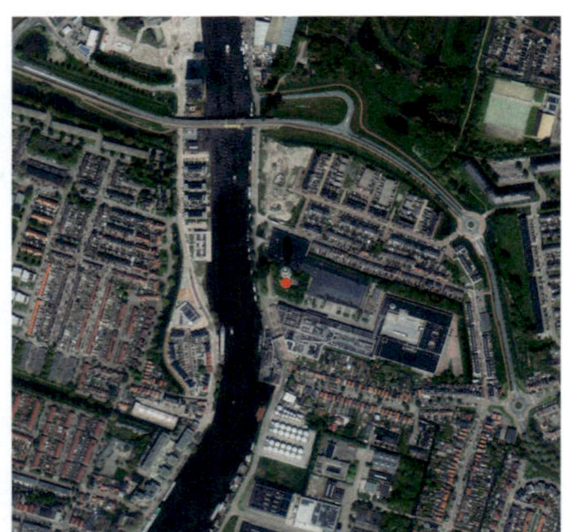

Cellnex Tower Wormer
Eenhoornweg 1
1531 ME Wormer

gerkens
DSV

HPE nimble (Google)
Tussenweg 16
1775 RK Middenmeer

Microsoft Amsterdam Data Center
Agriport 601
1775 TK Middenmeer

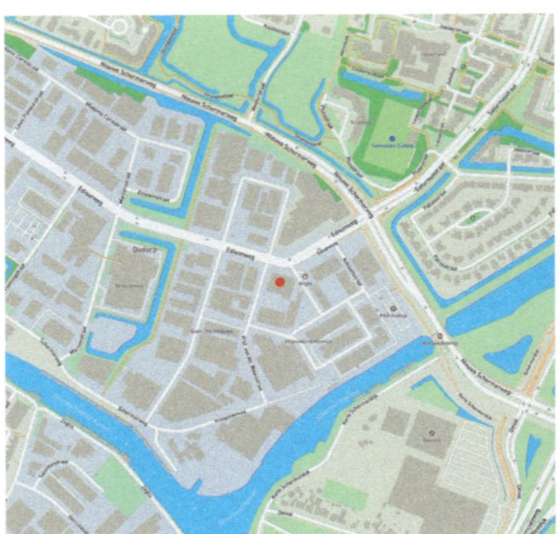

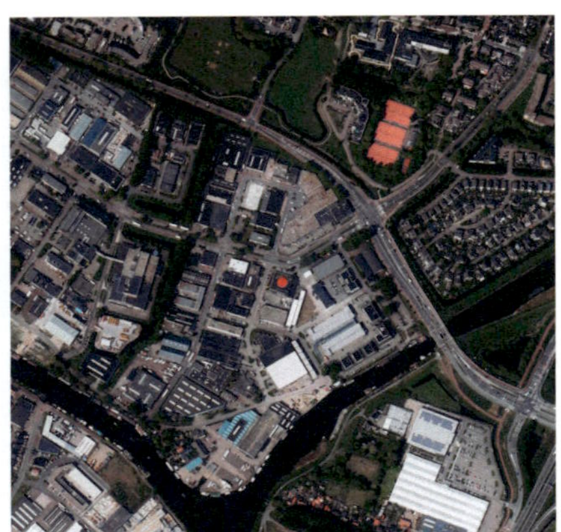

Datacenter Noord-Holland
Professor van der Waalsstraat 1
1821 BT Alkmaar

Iron Mountain AMS1
J. W. Lucasweg 35
2031 BE Haarlem

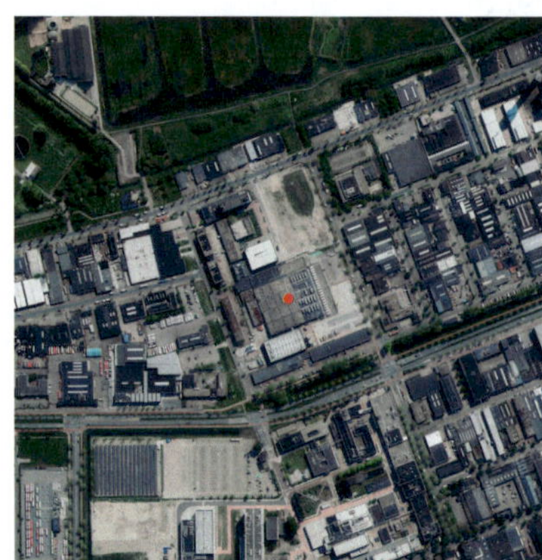

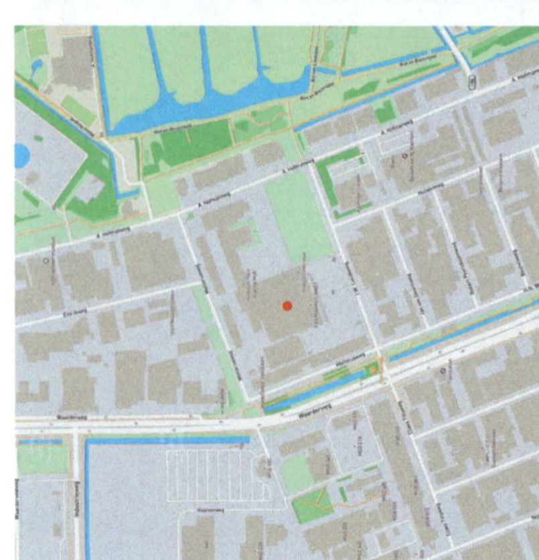

IRON MOUNTAIN
IRON MOUNTAIN

Cellnex Tower Haarlem
A. Hofmanweg 1
2031 BH Haarlem

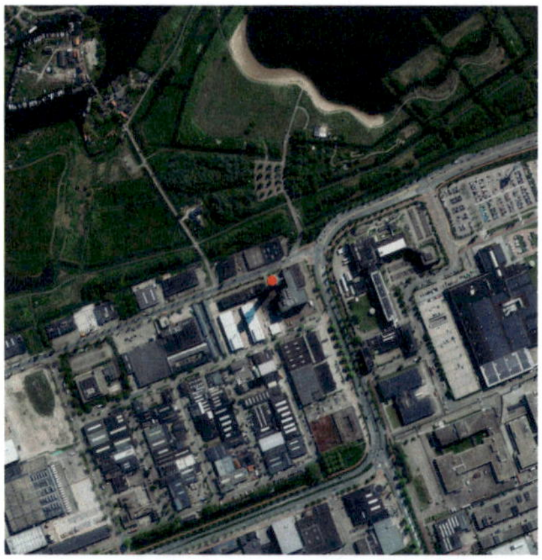

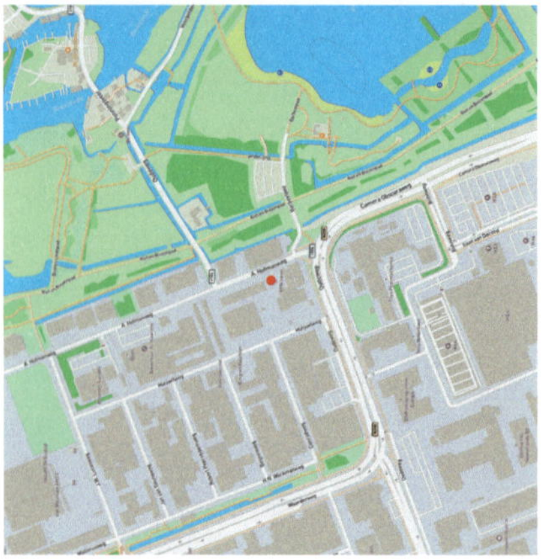

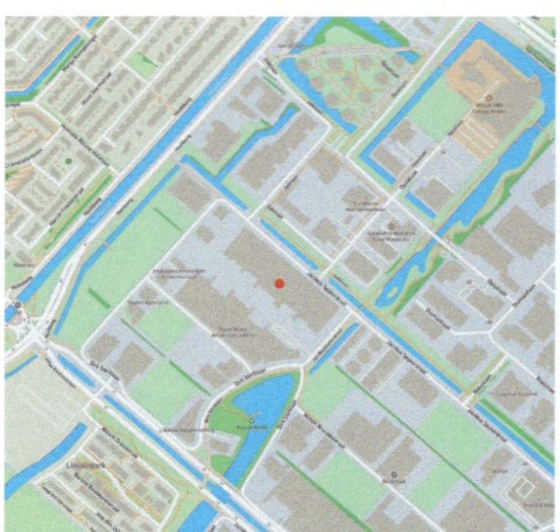

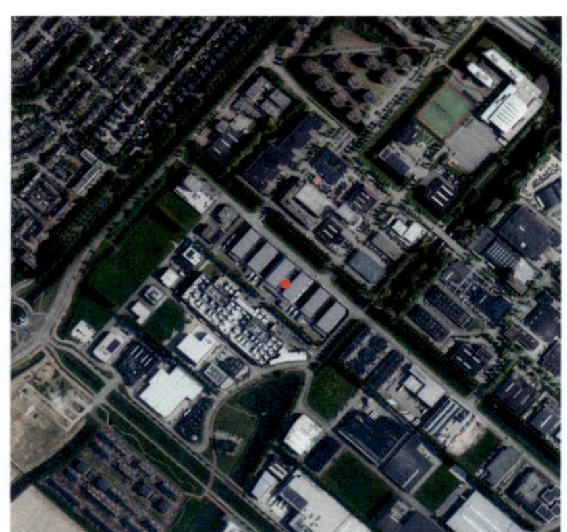

Digital Realty AMS16
De President
Jan Wijsmullerdreef 4
2132 PW Hoofddorp

BRUSH

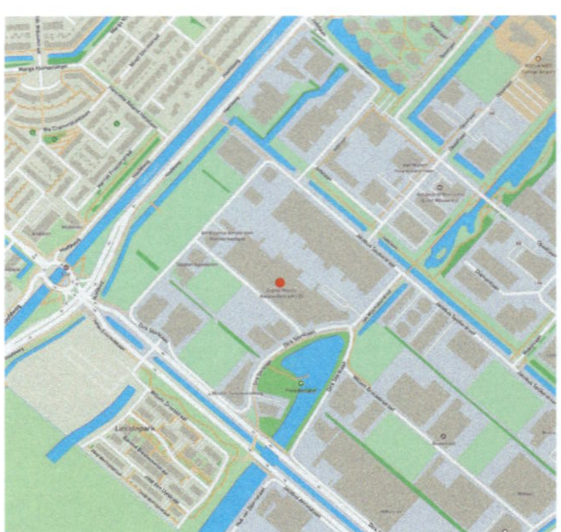

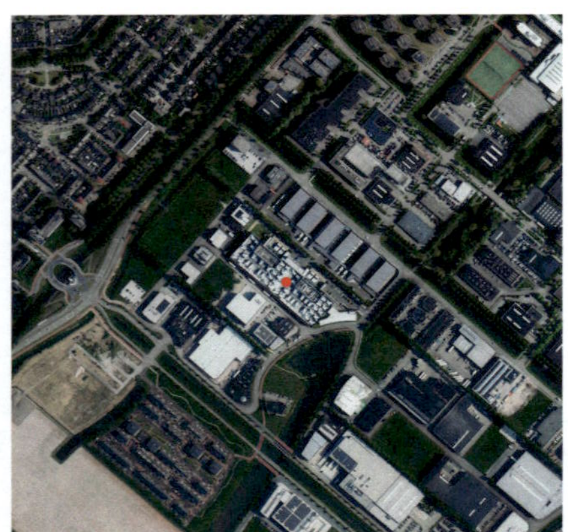

Digital Realty AMS15
De President
Jan Wijsmullerdreef 10
2132 PW Hoofddorp

Dataplace Amsterdam
Parellaan 9
2132 WS Hoofddorp

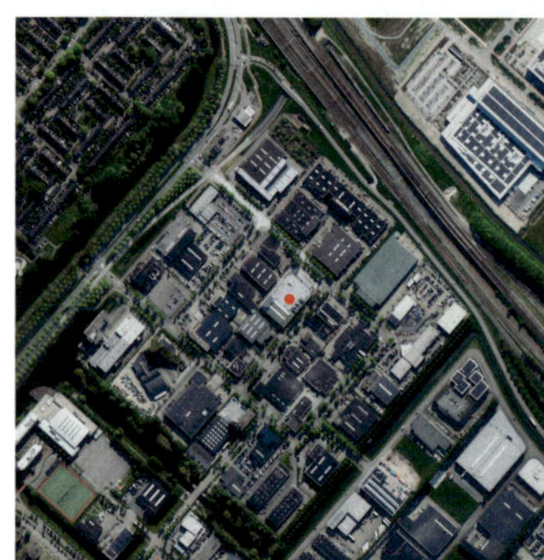

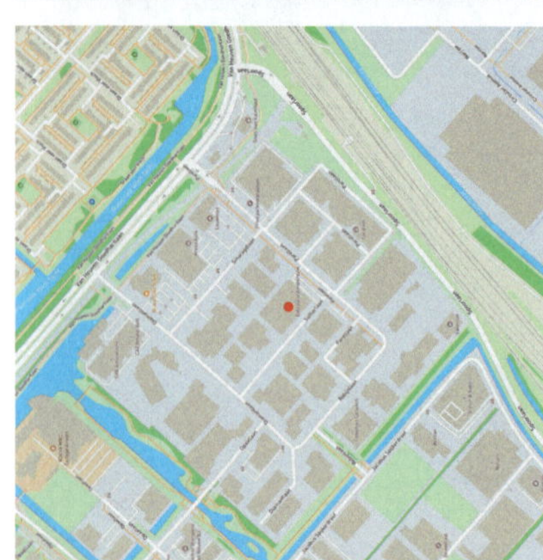

Cellnex Tower Alphen aan den Rijn
Energieweg 4
2404 HE Alphen aan den Rijn

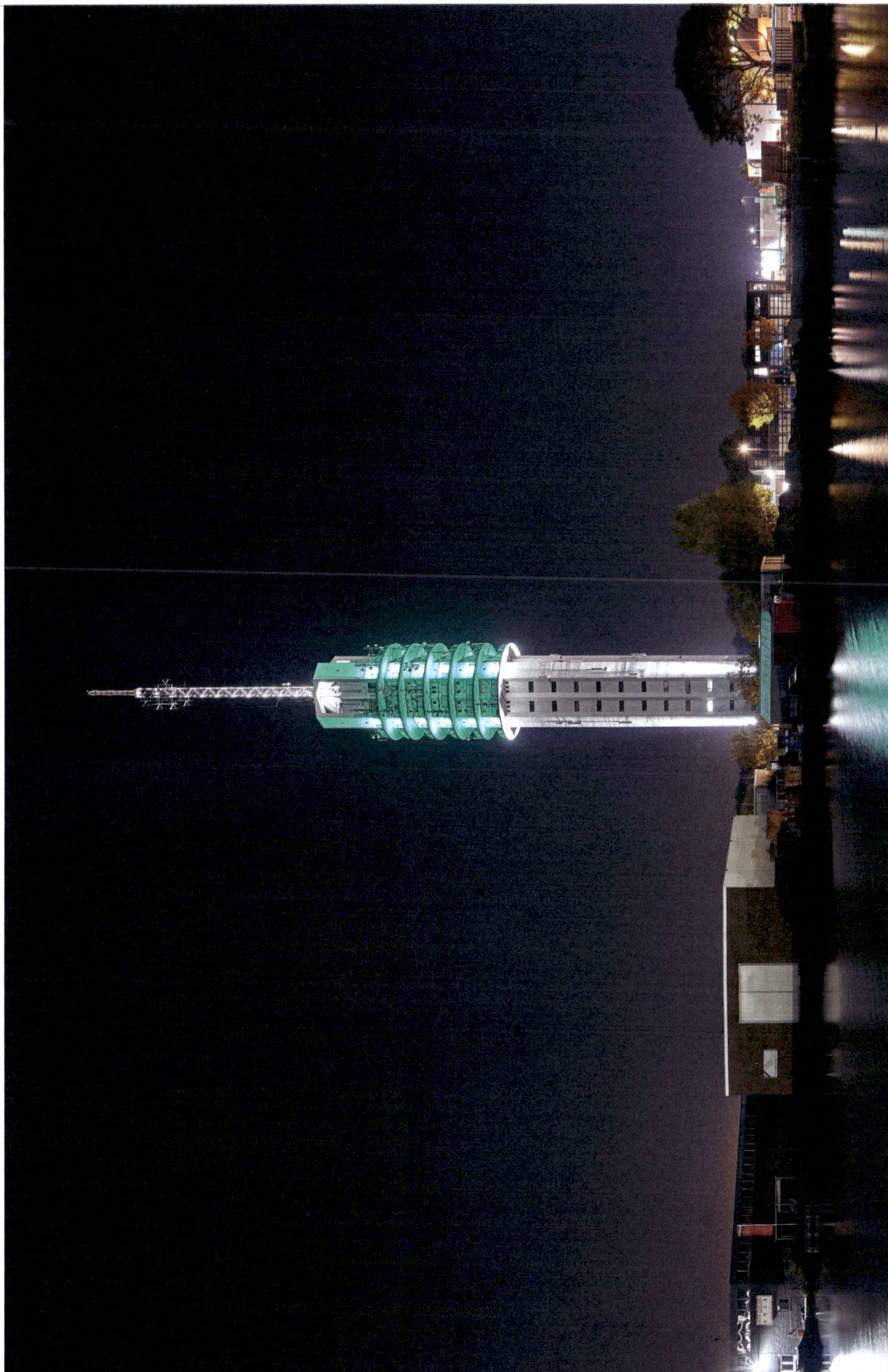

ColoHouse
Moezel 3–5
2491 CV Den Haag

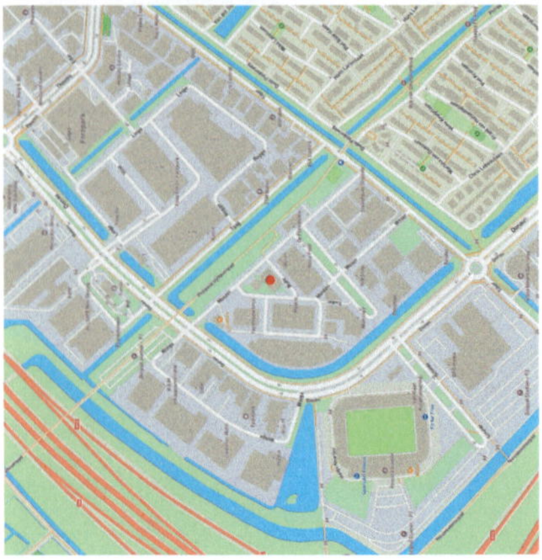

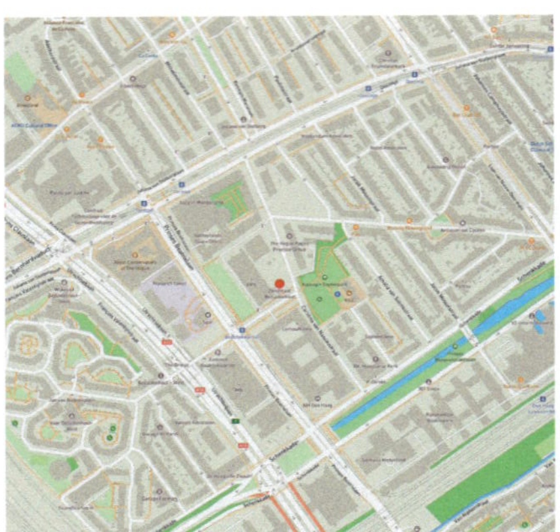

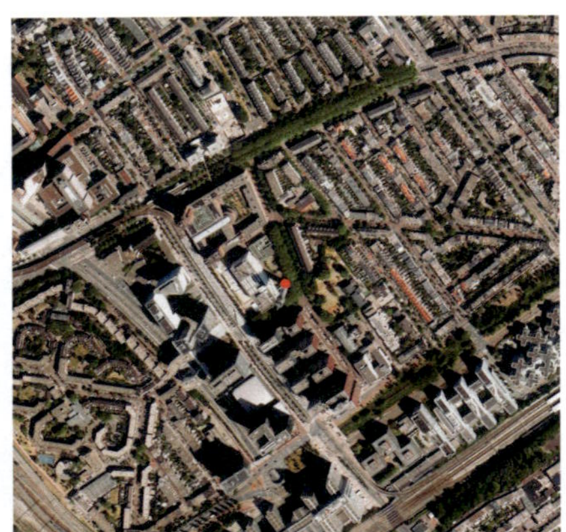

Cellnex Tower Den Haag
Louise Henriëttestraat 385
2595 BK Den Haag

NorthC Datacenter Delft
Heertjeslaan 1
2629 JG Delft

Dataplace Rotterdam
Van Coulsterweg 6
2952 CB Alblasserdam

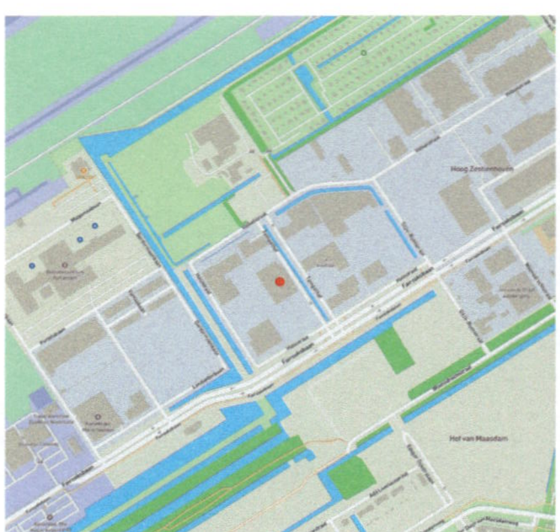

NorthC Datacenter Rotterdam
Zestienhoven
Tempelhof 5 – 11
3045 PV Rotterdam

Cellnex Tower Rotterdam
Anthony Fokkerweg 49
3088 GE Rotterdam

NorthC Datacenter Rotterdam
Waalhaven
Anthony Fokkerweg 40
3088 GG Rotterdam

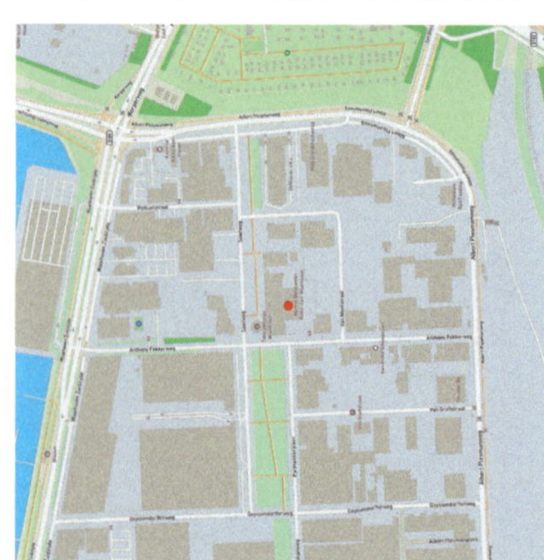

kpn

Data Facilities
Boyleweg 2
3208 KA Spijkenisse

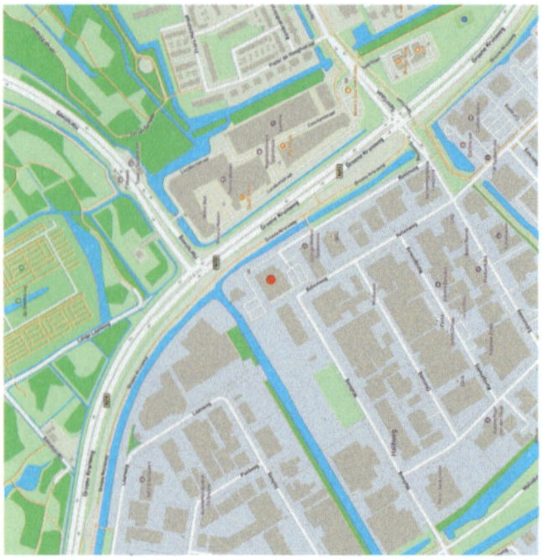

Boyleweg
2

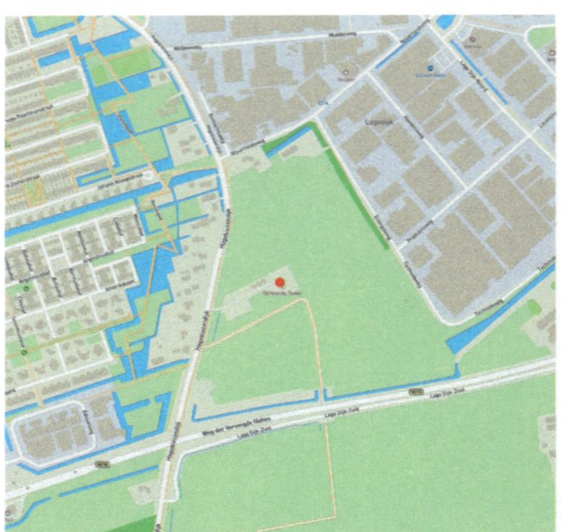

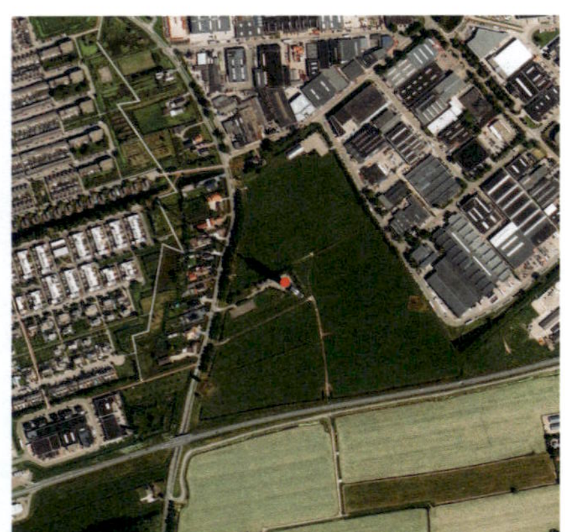

Cellnex Tower Ijsselstein
Gerbrandytoren
Hogebiezendijk 21
3401 RS IJsselstein

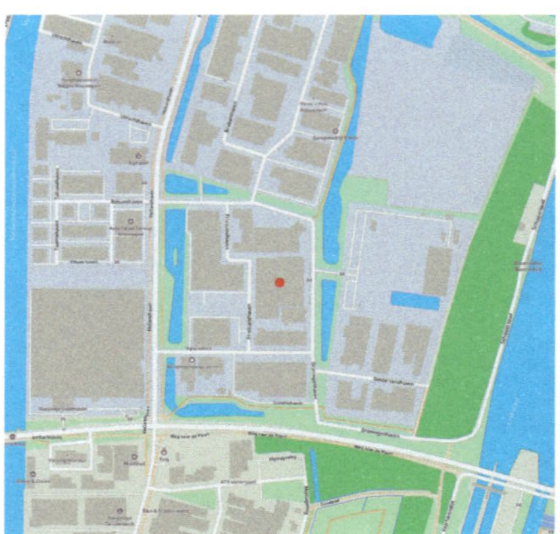

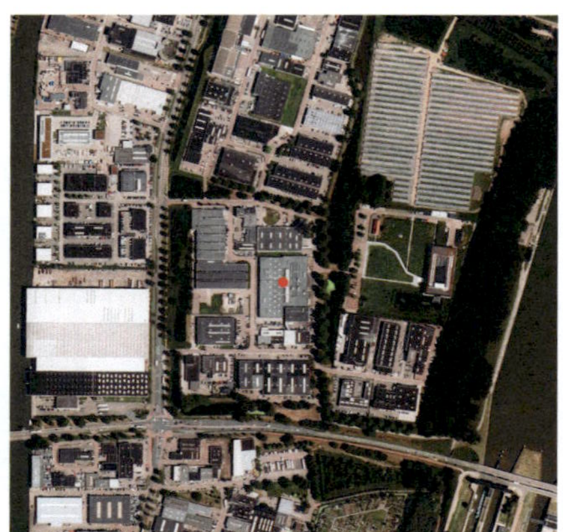

NorthC Datacenter Nieuwegein
Groningenhaven 26
3433 PE Nieuwegein

Switch Datacenter Group
Polanerbaan 1
3447 GN Woerden

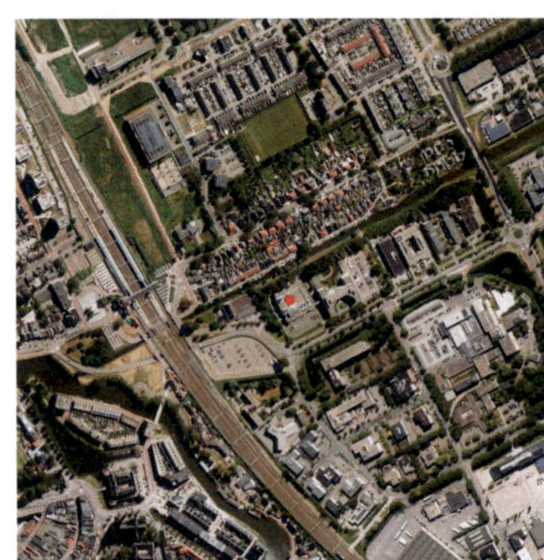

Dataplace Utrecht
Kon. Wilhelminaweg 471
3737 BE Groenekan

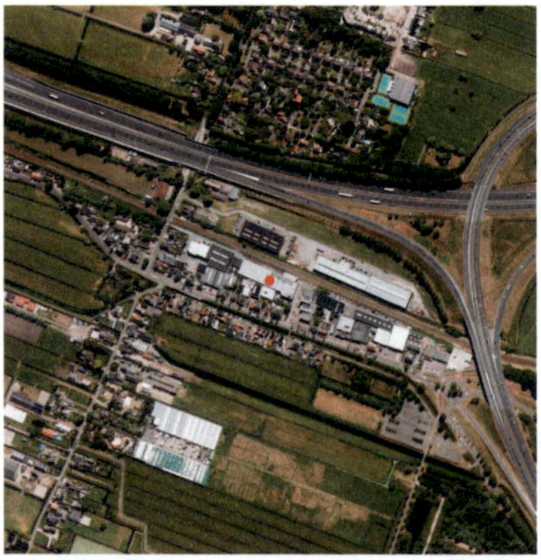

Cellnex Tower Goes
Cornelisweg 10
4463 AK Goes

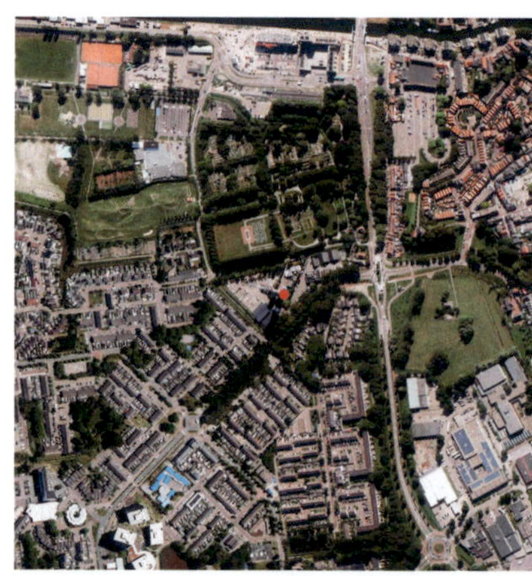

Dataplace Nedzone
Drukkerij 6
4651 SL Steenbergen

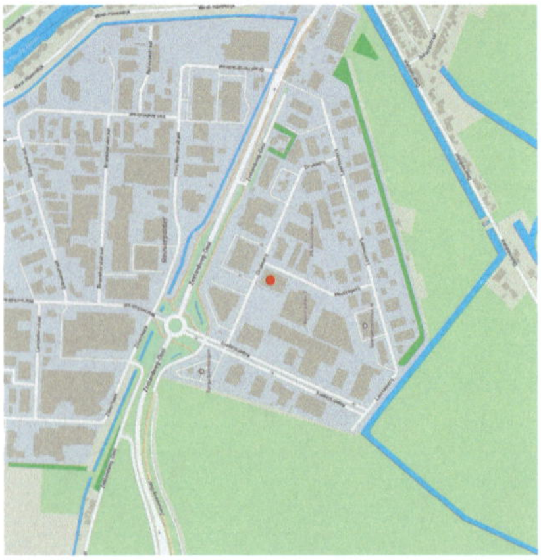

data place
Ingang

Colt Roosendaal
Argonweg 9
4706 NR Roosendaal

Cellnex Tower Roosendaal
Melis Stokelaan 25
4707 HP Roosendaal

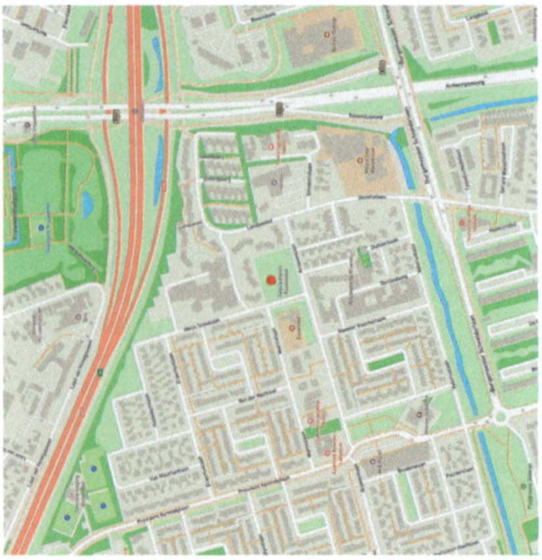

Global-e Datacenter
Haansbergseweg 12
5121 LJ Rijen

GLOBAL-e

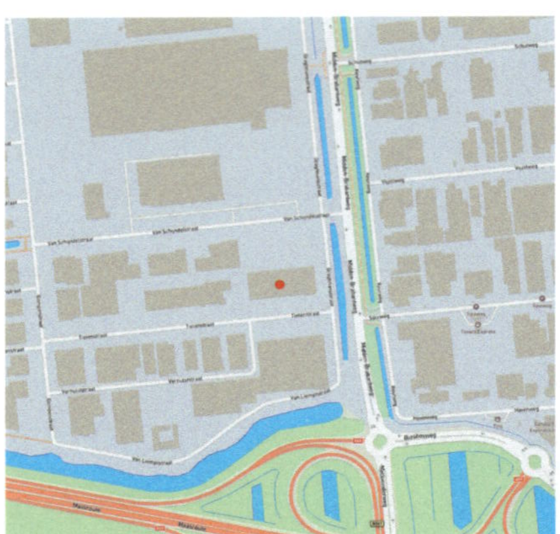

Eurofiber Datacenter Brabant
(Dataplace Brabant)
Gragtmansstraat 1B
Industrieterrein Haven 7
5145 RA Waalwijk

eurofiber

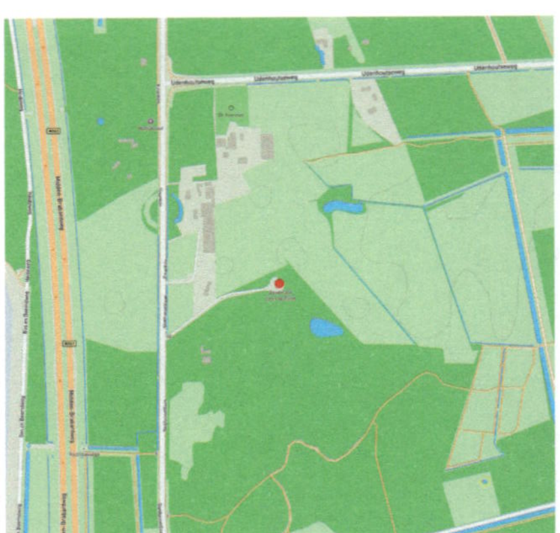

Cellnex Tower Loon op Zand
Finantiën 5
5175 NW Loon op Zand

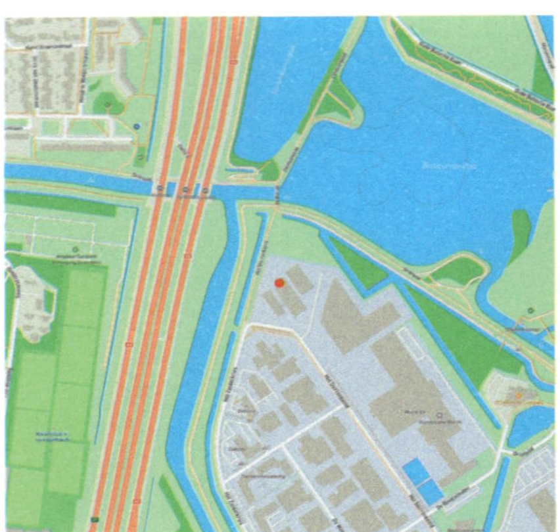

Interconnect Services
Het Sterrenbeeld 55
5215 MK 's-Hertogenbosch

INTERCONNECT

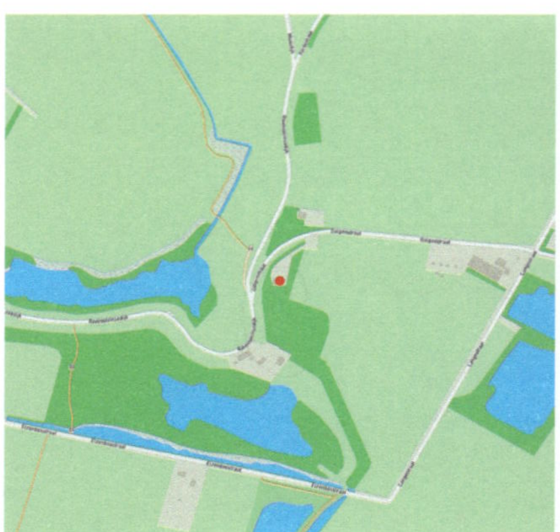

Cellnex Tower Haren
Ravensteinsedijk 2A
5368 LB Megen

Cellnex Tower De Mortel
Hemelsbleekweg 15
5425 PB De Mortel

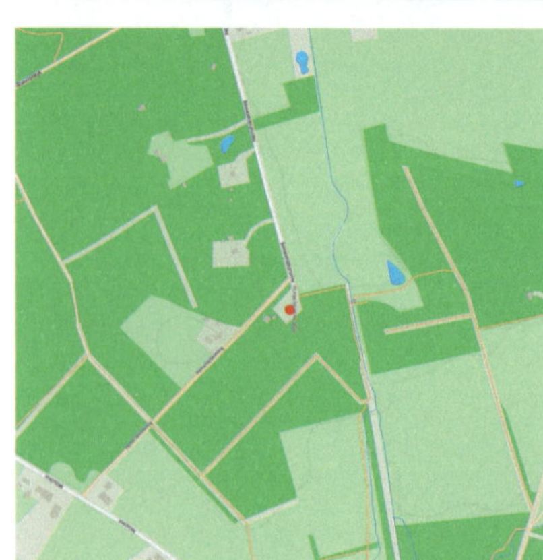

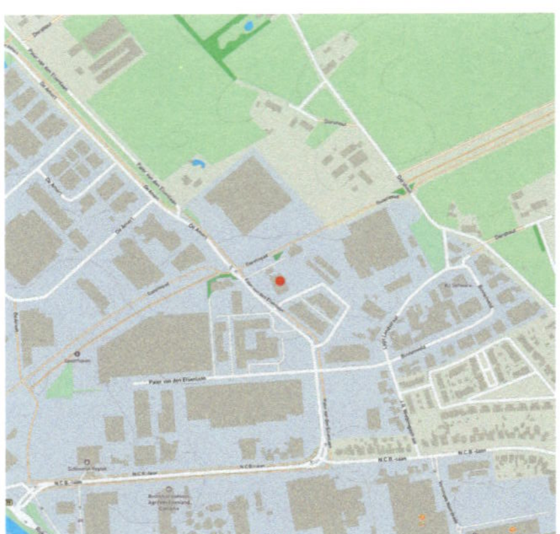

e-Quest Datacenter
Pater van den Elsenlaan 30
5462 GG Veghel

e-Quest
30

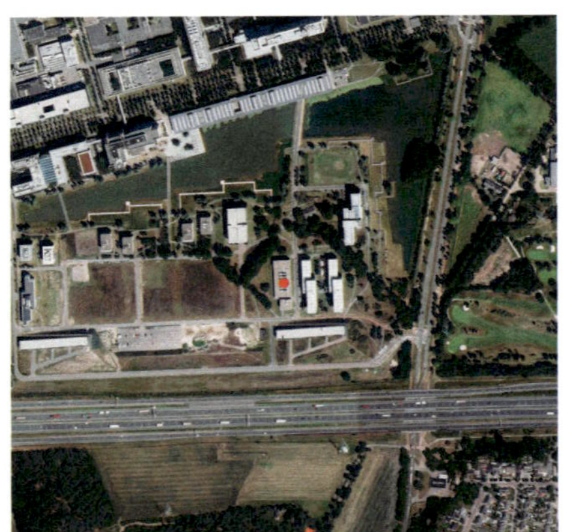

NorthC Datacenter Eindhoven
High Tech Campus 53
5656 AE Eindhoven

NorthC
NorthC
NorthC

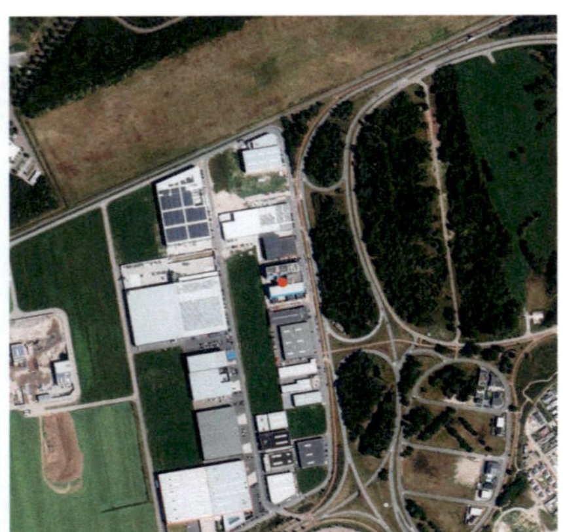

Interconnect Services
Park Forum 1041
5657 HJ Eindhoven

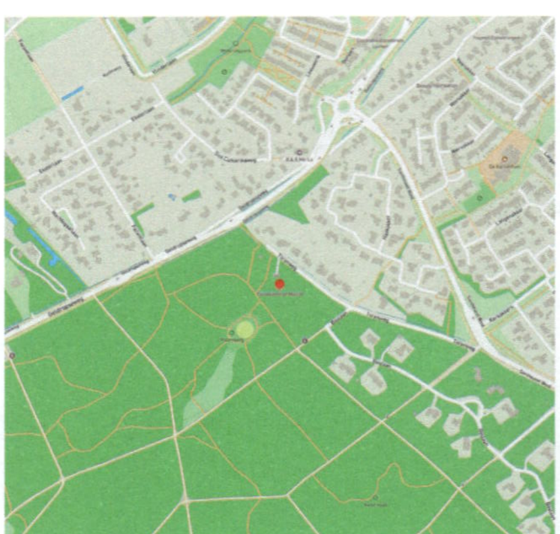

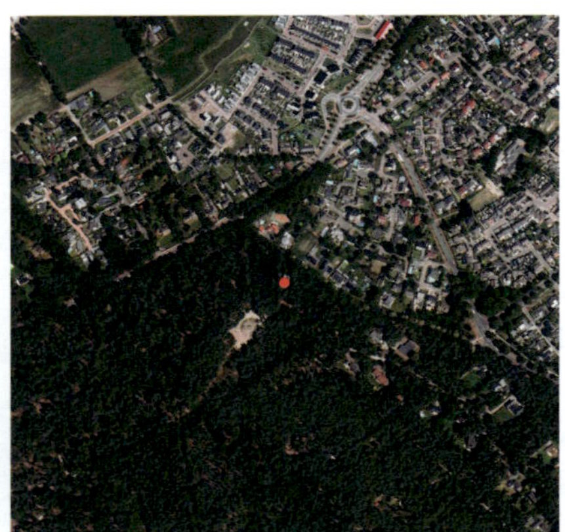

Cellnex Tower Mierlo
Torenweg 1
5731 CJ Mierlo

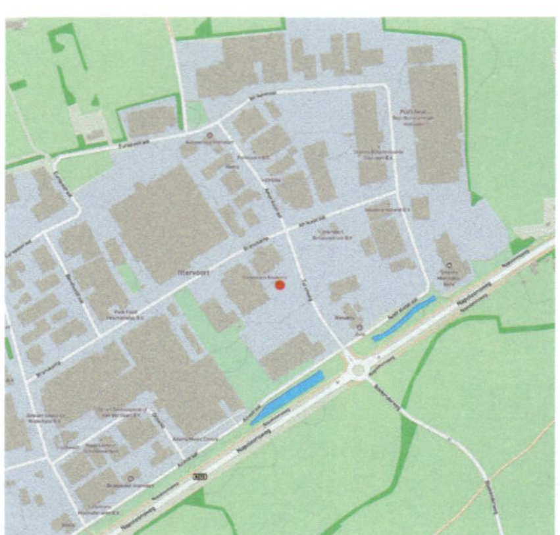

Cellnex Tower Ittervoort
Branskamp 16
6014 CB Ittervoort

Cellnex Tower Roermond
Op de Meuleberg 7
6041 NK Roermond

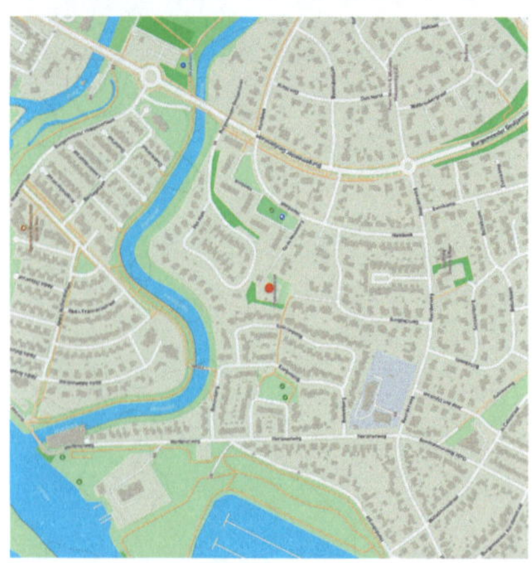

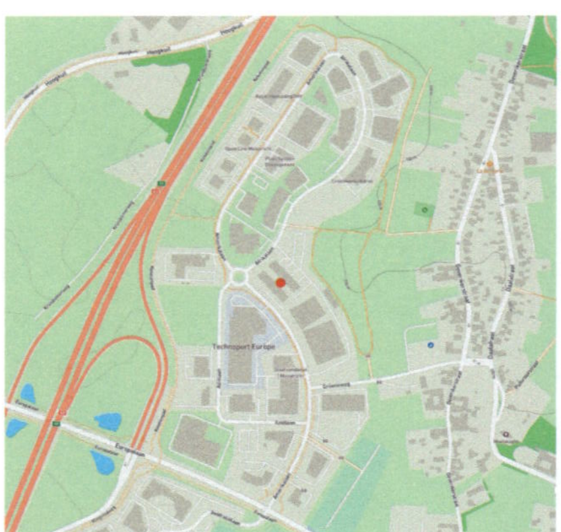

ENGIE Services Zuid
Cofely Datacenter
Amerikalaan 35
6199 AE Maastricht Airport

engie

Cellnex Tower Maastricht
Trichterbaan 170
6215 XZ Maastricht

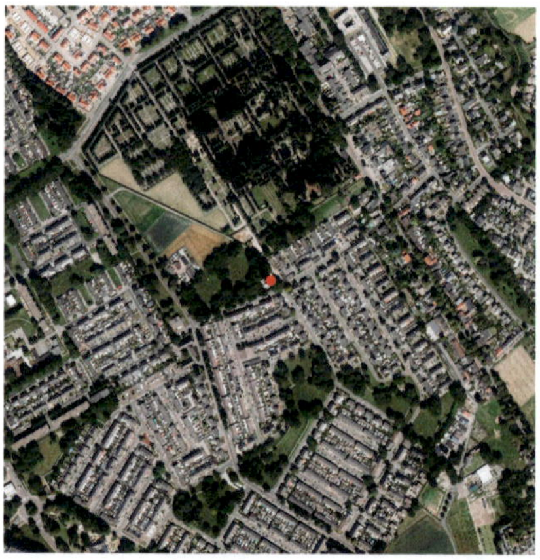

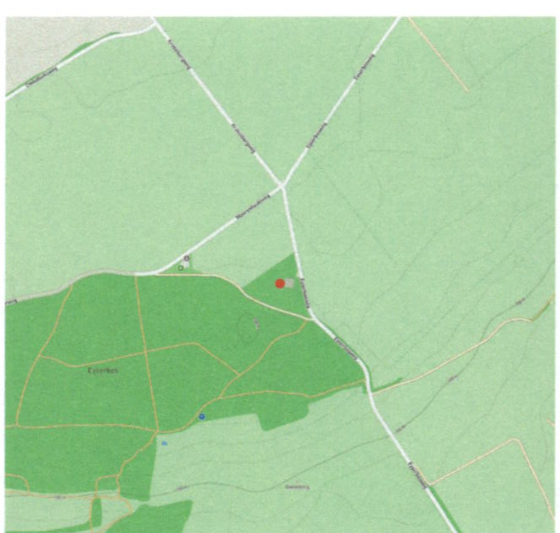

Cellnex Tower Eys
Eyserbosweg 27
6287 NE Eys

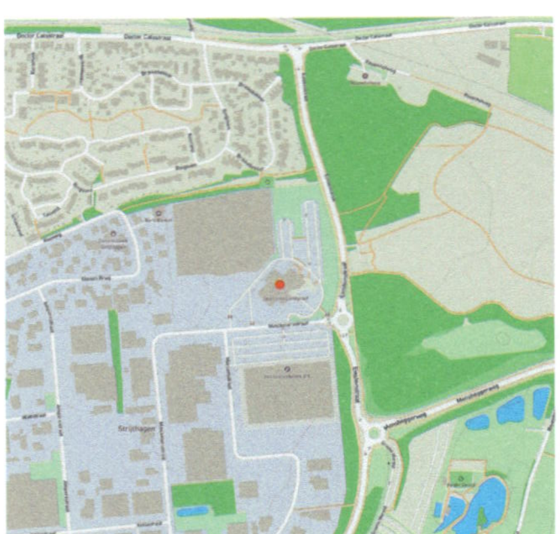

ENGIE Services
(Open Line)
Minckelersstraat 2
6372 PP Landgraaf

Dataplace Arnhem
Tivolilaan 251
6824 BV Arnhem

Serverius IT Infrastructure
Laan van de Ram 39
7324 BW Apeldoorn

serverius
IT INFRASTRUCTURE

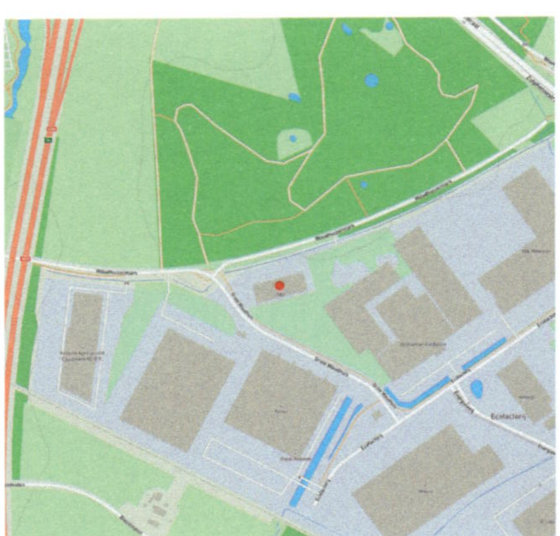

ITB2 Datacenters
Grote Woudhuis 5
7325 WM Apeldoorn

ITB
Datacenters

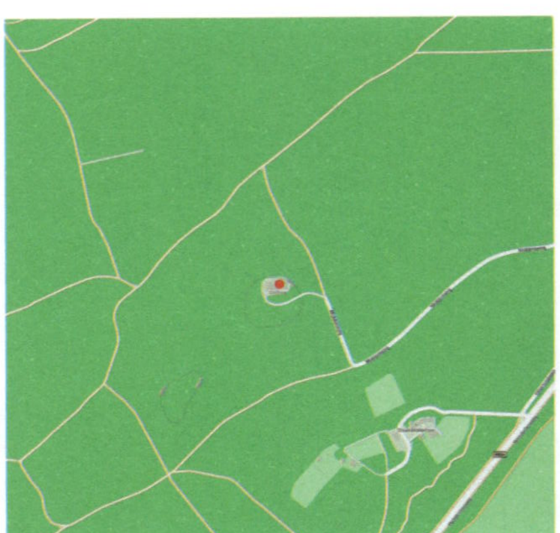

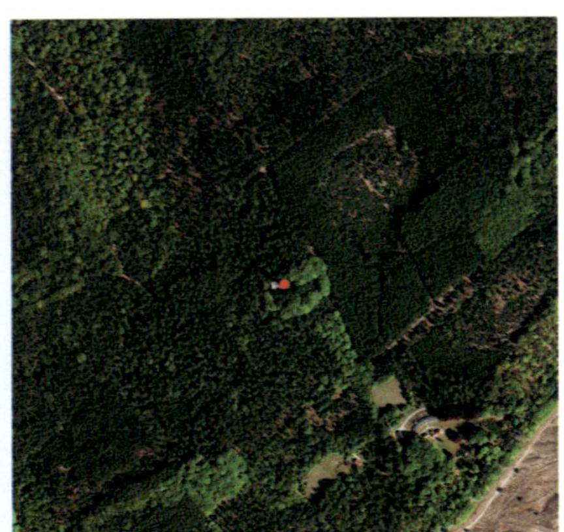

Cellnex Tower Ugchelen
Otterloseweg 110
7339 GZ Ugchelen

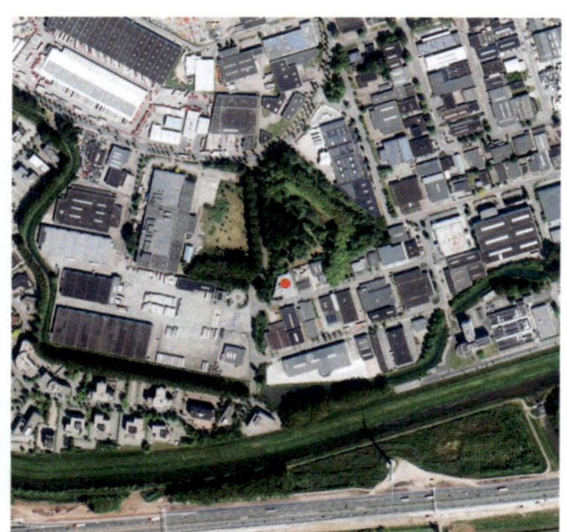

ITB2 Datacenters
Herfordstraat 18
7418 EX Deventer

Cellnex Tower Markelo
Larenseweg 58
7475 PX Markelo

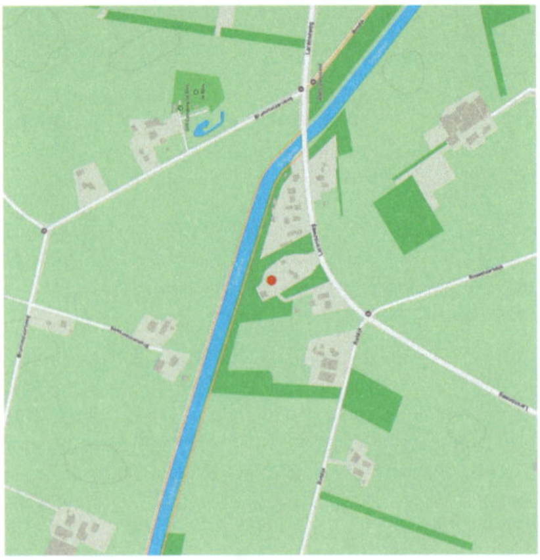

Equinix EN1
Auke Vleerstraat 1
7521 PE Enschede

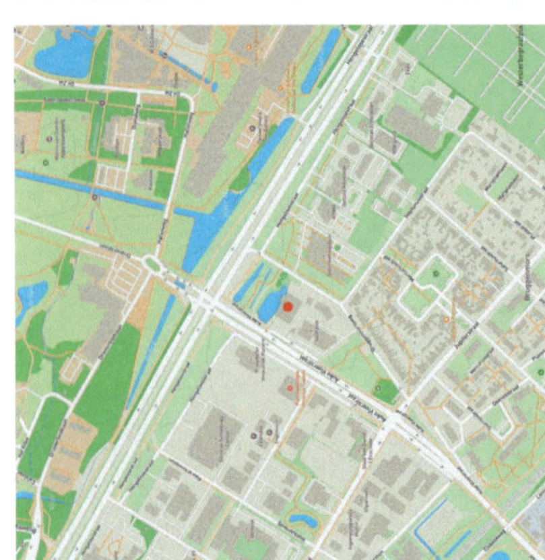

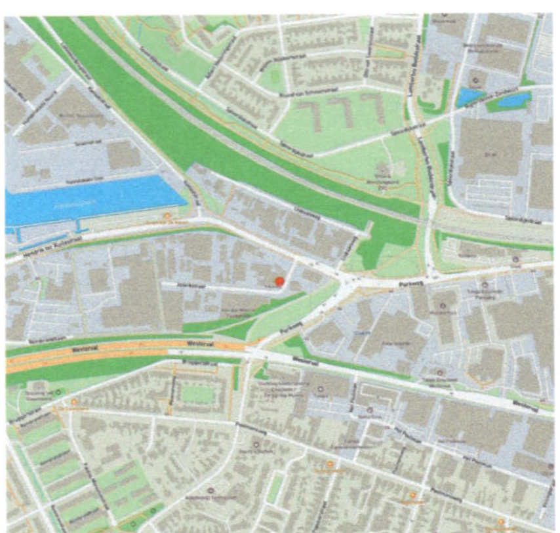

InterDC
Josinkstraat 28
7547 AB Enschede

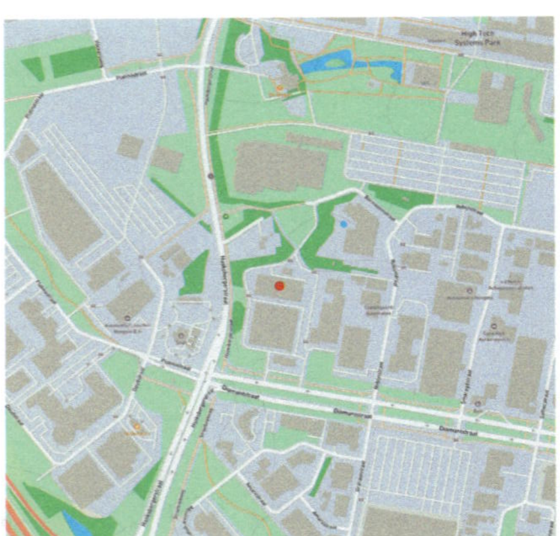

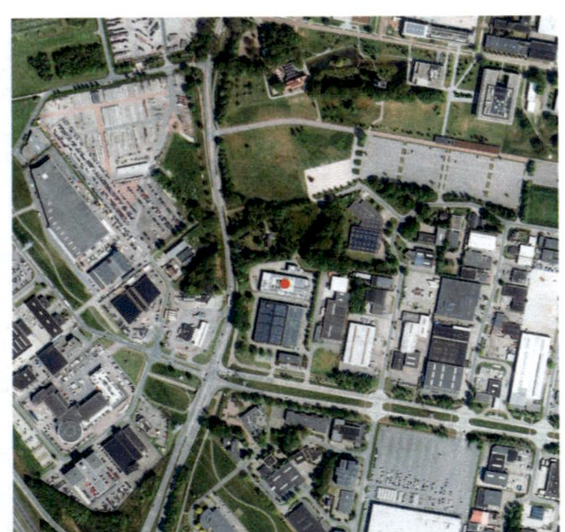

Previder PDC2
Barnsteenstraat 15
7554 TC Hengelo

previder
it starts here
10

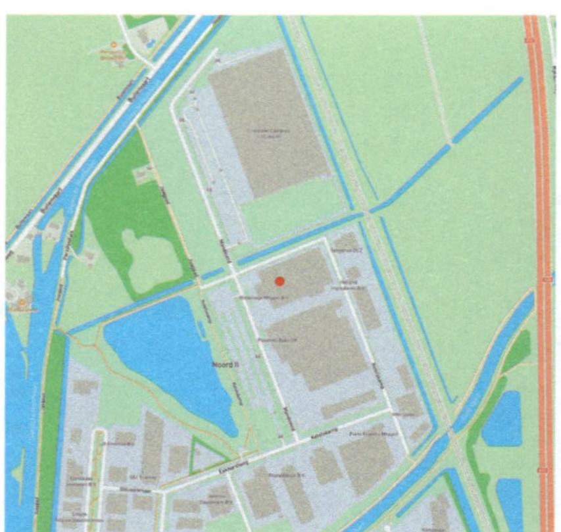

Serverius IT Infrastructure
Ketelskamp 10
7942 KG Meppel

serverius
DATACENTERS

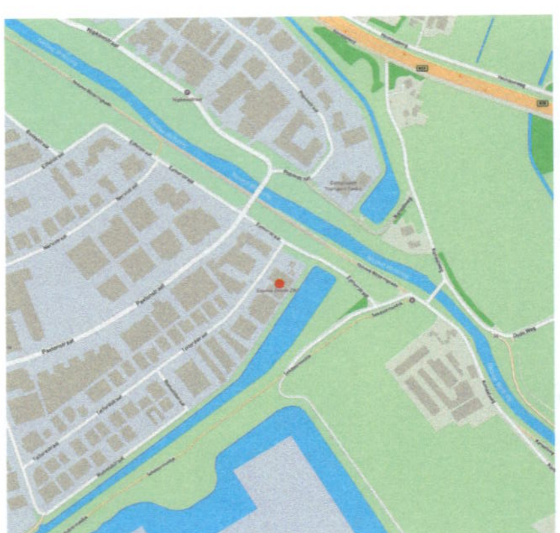

Equinix ZW1
Telfordstraat 3
8013 RL Zwolle

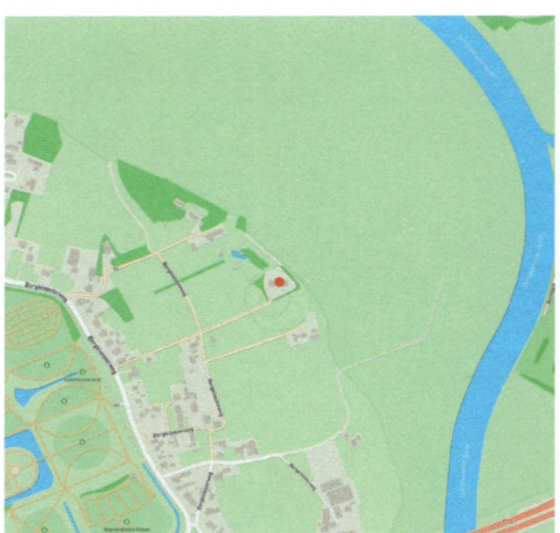

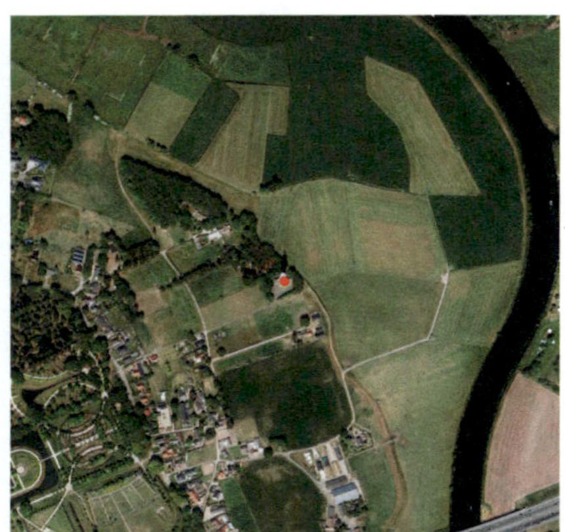

Cellnex Tower Zwolle
Bergkloosterweg 62
8034 PP Zwolle

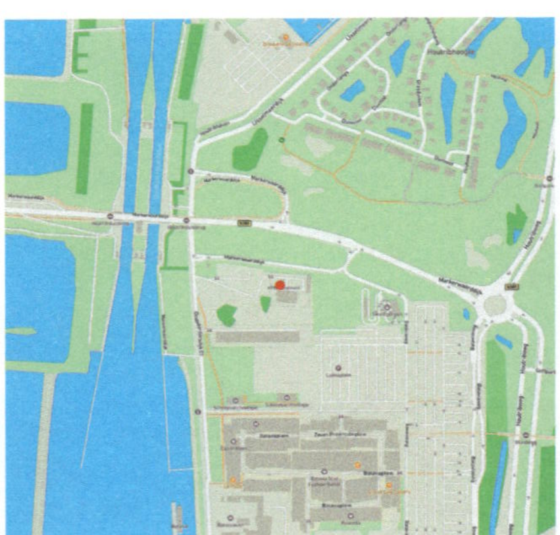

Cellnex Tower Lelystad
Oostvaardersdijk 1 – 3
8244 PA Lelystad

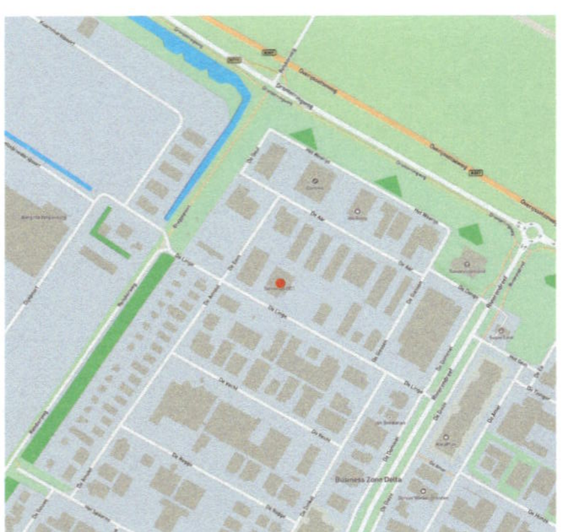

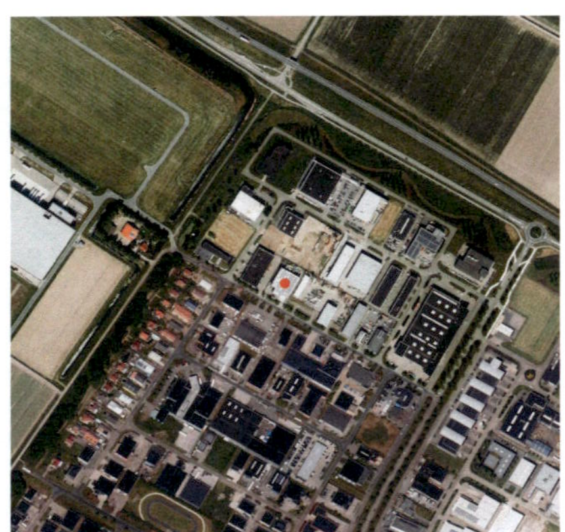

Serverius IT Infrastructure
De Linge 26
8253 PJ Dronten

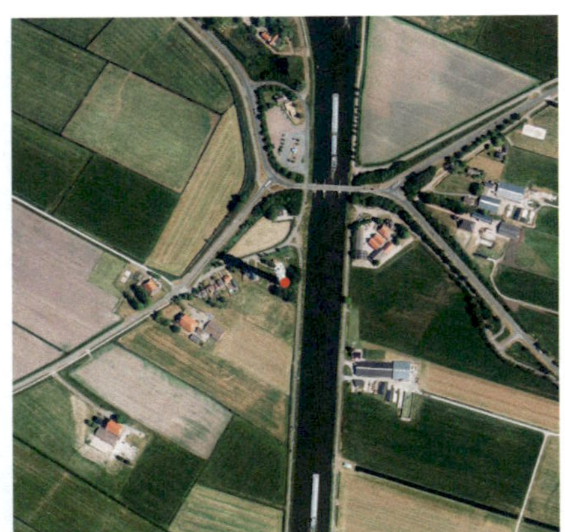

Cellnex Tower Tjerkgaast
Gaestdyk 3
8522 MV Tjerkgaast

368

DataCenter Fryslân 2
Archimedesweg 3
8912 AK Leeuwarden

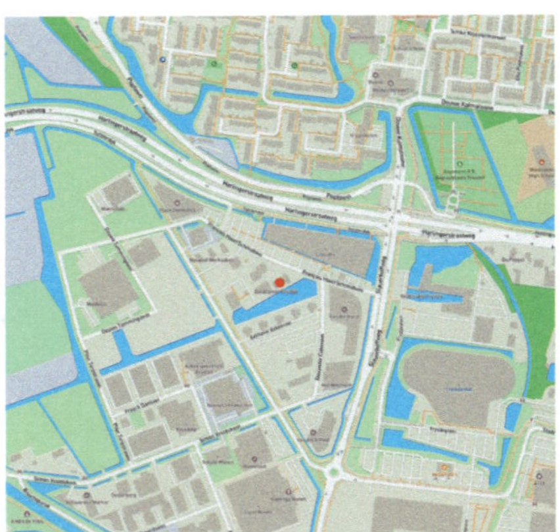

DataCenter Fryslân 1
François HaverSchmidtwei 3
8914 BC Leeuwarden

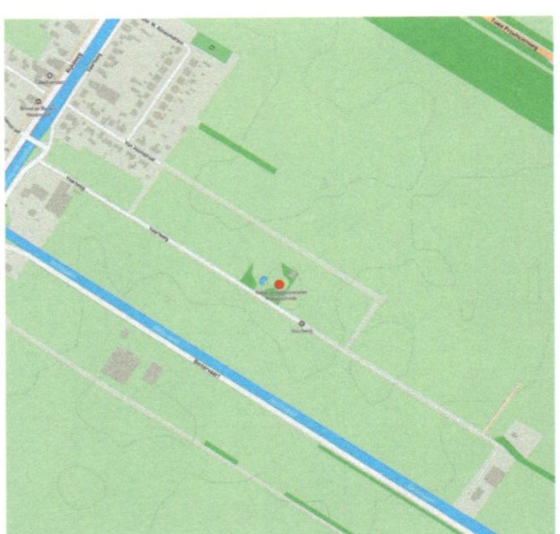

Cellnex Tower Hoogersmilde
Zendstation Smilde
Van Veenstraat 26
9423 VA Hoogersmilde

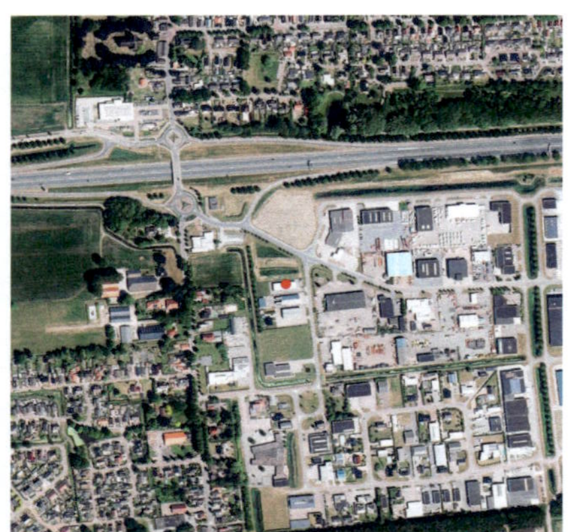

Datacenter Groningen
Beneluxweg 4
9636 HV Zuidbroek

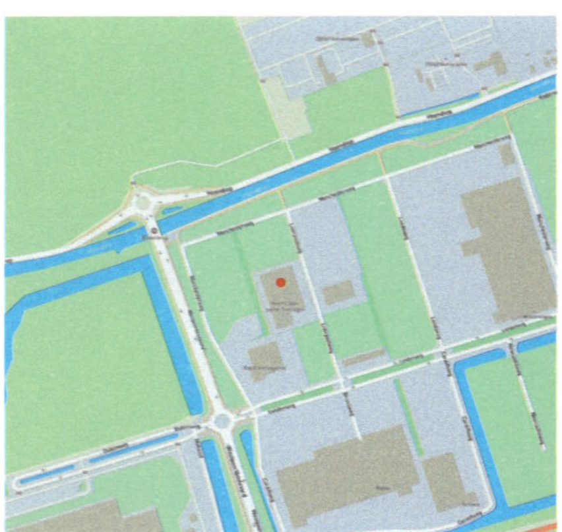

NorthC Datacenter Groningen
Liverpoolweg 10
9744 TW Groningen

QTS Groningen
Zernikelaan 16
9747 AA Groningen

Bytesnet Groningen
De Bunders 1
9747 AB Groningen

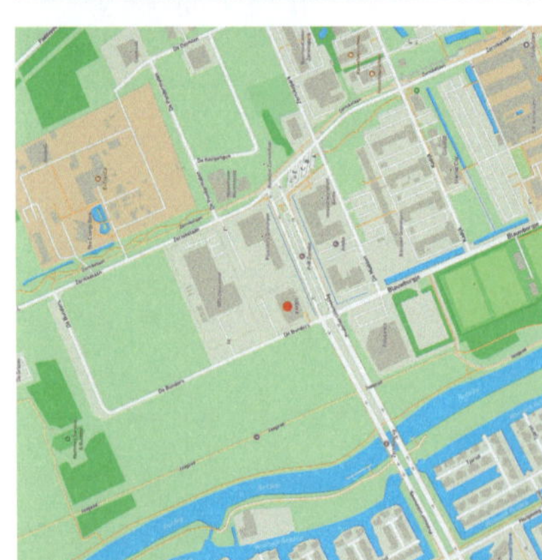

bytesnet

Smitsborg
Nettelbosje 1
9747 AJ Groningen

Appendix

Technical Glossary

Amsterdam Internet Exchange
The Amsterdam Internet Exchange (AMS-IX) is one of the largest Internet exchange points (IXPs) in the world. It is located in Amsterdam Science Park. Established in the early 1990s, AMS-IX is a nonprofit, neutral and independent peering point.

Blockchain
A blockchain is a decentralized and distributed digital ledger that records transactions across multiple computers in a way that these entries cannot be altered retroactively without modifying all subsequent blocks of the chain. Data centres frequently host large mining operations that require significant computational resources to solve cryptographic puzzles and validate transactions.

Carrier-Neutral
The term carrier-neutral (or: network-neutral) refers to a facility or provider that allows interconnection with multiple telecommunications carriers or network service providers, without favouring any specific company. Carrier-neutral data centres are not tied to any one service provider providing diversity and flexibility for the client seeking service.

Cloud
The 'cloud' as a representation of the Internet comes originally from the field of telecommunications and network diagrams. Here, these diagrams used the cloud symbol as a way to simplify the underlying infrastructure of telephone networks. According to Wikipedia, use of the cloud metaphor to denote virtualized services traces back to 1994, when it was used by General Magic to 'think of its electronic meeting place as the cloud'.

Cloud Computing
Cloud computing is the delivery of computing services over the Internet. Rather than keeping files on a local storage device, cloud-based storage allows for remote information management and on-demand data access. The first implementation of modern cloud computing was launched in 2002 by Amazon.

Colocation Data Centre
Colocation, or collocation, refers to a data centre facility in which multiple customers locate their servers and networking equipment. These customers rent space, power and bandwidth from the colocation provider, which owns and manages the facility. Big international colocation providers include companies like Equinix, Iron Mountain and Digital Realty.

Edge Computing
Edge computing refers to a distributed form of computing where data processing is performed close to the data source or at the edge of the network, rather than in the centralized cluster of servers. This emerging computing paradigm is expected to improve response times and save bandwidth.

High Performance Computing
In order to solve complex and resource-intensive computing problems, High Performance Computing (HPC) helps to aggregate the resources of multiple servers in a data centre. By means of supercomputing and simultaneous processing, HPC data centres allow to significantly reduce the necessary computation time.

Hot and Cold Aisle
The cold aisle is the side of a data centre corridor where the fronts of server racks face each other. The hot aisle on the other hand is located adjacent to the cold aisle, where the warm exhaust air from the server ventilation congregates (hence the name). The hot aisle versus cold aisle arrangement is a common data centre cooling strategy that helps optimize airflow and temperature management.

Hyperscale Data Centre
Hyperscale data centres are large-scale data repositories designed to process the massive demands of cloud computing operations from one single tenant. These types of data centres are built for maximum efficiency and frequently associated with globally operating Internet companies such as Google, Apple, Amazon, Meta and Microsoft.

Internet Service Provider
An Internet service provider (ISP) is a company or organization that provides Internet access to customers. The first ISP was Telenet, introduced in 1974.

Latency
Network latency refers to the time delay that data takes to transfer across a network. A long delay indicates high latency, while faster response times have low latency. That's why data centres are usually situated in proximity to big network hubs like the Amsterdam Internet Exchange (AMS-IX). Latency is usually measured in milliseconds (ms) or microseconds (µs).

Nitrogen
Nitrogen is related to data centres primarily through its use in advanced cooling and fire suppression systems. In some high-performance computing applications, liquid nitrogen helps to protect processors from overheating through cooling and to create a dry atmosphere by preventing condensation on electronic components.

Redundancy
Data redundancy refers to the duplication of critical data components to ensure continued operation and minimize the risk of service interruptions. This key design principle is employed to mitigate the impact of hardware failures and software crashes. It means essentially storing the same information in multiple locations to enhance data reliability and availability.

Scalability
In the context of cloud computing, scalability relates to the ability of a data centre to increase or decrease its resources and capabilities to meet changing market demands. It involves efficiently handling varying workloads, growing user bases, and expanding data without compromising performance or reliability. Horizontal scalability involves adding more machines or nodes to the data centre network and vertical scalability refers to amplifying computing power (CPU, RAM, storage) to existing machines.

Streaming
Growing streaming numbers have a significant impact on data centres and their services. Excessive binge-watching results in an increased demand for high-bandwidth connections and storage capacities, and in a fierce rat race for cooling and power. One hour of streaming is estimated to consume 3.2 kg of CO_2.

Supercomputing
Supercomputers are extremely powerful computers designed to perform complex calculations at high speeds. These machines are capable of processing vast amounts of data and executing trillions of calculations per second, far surpassing the capabilities of standard computers. Europe's most famous supercomputer is probably MareNostrum from the Barcelona Supercomputing Center.

Tiers
Data centre tiers are defined levels of resiliency and redundancy for IT facility infrastructure. The classification ranks facilities from 1 to 4, with 1 being the worst and 4 the best-performing level. Data centre facilities receive their ranking from Uptime Institute, a Seattle-based digital infrastructure authority.

Uninterruptible Power Supply
Uninterruptible Power Supply (UPS) is a critical component of the infrastructure that provides backup power during electrical outages, fluctuations or interruptions. Data centres house a large number of servers, networking equipment and storage systems that are sensitive to power disruptions. A UPS ensures a continuous power supply, allowing data centre operations to continue without interruption and protecting against potential data loss or system damage.

Uptime and Downtime
Uptime and downtime are terms used to measure the reliability and availability of IT systems, but also data centres. They represent the periods during which a system is operational (uptime) or not operational (downtime). Typically, they are expressed as a percentage and calculated by dividing the total operational time by the total time in a given period.

Waste Heat
Waste heat is the thermal energy generated as a byproduct during uptime. Data centres house large numbers of servers and other computing infrastructure that work continuously to process and store information. As these devices operate, they consume electrical power and give off heat. For large-scale data centres, the waste heat can be so vast that they can provide heating to thousands of homes and businesses.

White Space
In the context of data centres, the white space refers to the areas within a building that are allocated for IT equipment deployment, such as servers, storage systems, networking devices and other computing hardware. It is in fact, the vacant space available to customers for lease – hence the term white.

Equinix AM7 Data Center, Amsterdam
Alticom Datacenter, Roosendaal
Digital Realty
Amsterdam Data Tower
Dataplace Rotterdam
Alticom Data
uinix AM5 Data Center, Amsterdam
Alticom Datacenter (Gerbrandytoren), IJsselstein
Serverius SD
center, Hilversum
Equinix AM8 Data Center, Amsterdam
KPN Datace

Hoofddorp
Iron Mountain AMS 1, Haarlem
Alticom Datacenter (ViaData Toren), Tjerkgaast
Equinix AM1 Data Center, Amsterdam
Equinix AM6 Data Center, Amsterdam
Equinix AM4, Amsterdam
Interxion AMS 8, Aalsmeerderbrug
Digital Realty Data Center, Amsterdam
Alticom Datacenter, Hoo

Contributors

Ramon Amaro is Senior Researcher in Digital Culture at Nieuwe Instituut in Rotterdam and tutor in Media Theory at the Royal Academy of Art in The Hague. An engineer and sociologist by training, Amaro's writings, research and artistic practice emerge at the intersections of Black study, digital culture, psychosocial study and the critique of computational reason. Amaro has worked as Lecturer, Art and Visual Cultures of the Global South at University College London, engineering program manager for the American Society of Mechanical Engineers and quality design engineer for General Motors. His recently published book, *The Black Technical Object: On Machine Learning and the Aspiration of Black Being* (Sternberg, 2023) contemplates the abstruse nature of programming and mathematics, and the deep incursion of racial hierarchy, to inspire alternative approaches to contemporary algorithmic practice.

Roel Backaert began his photography studies at Studenteninformatiesysteem Antwerpen (SisA) in Antwerp, and also studied at the Sint-Lukas Academie in Brussels. He lives and works in Amsterdam where he graduated from the Gerrit Rietveld Academie in 2004. Since then he has been travelling the world as an architectural photographer and pursued a career as an autonomous artist. The main focus in his work is the urban landscape from which he often plucks a peculiar object, building or cityscape out of its context to present it in compelling centralized compositions. Backaert works mostly at night with a 4 × 5 inch (10.2 × 12.7 cm) analogue camera using very long shutter-speeds that pull the subjects out of the darkness into an exceptionally colourful twilight.

Sheena Calvert is a philosopher, writer, designer and artist. Her PhD is in the philosophy of language and aesthetics. With a background in typography and book design, she has always been immersed in questions about the materiality of language, and its implications for how we form knowledge. In her work as a practitioner and educator, her practice engages with philosophy and politics, and is located at the intersection between language, technology and ethics. She is currently focussing on the shifting relationships between these as a result of the emergent interfaces between language and human / nonhuman agents.

Annet Dekker is Assistant Professor, Cultural Analysis, and Archival and Information Studies at the University of Amsterdam. She is also Visiting Professor and Co-director of the Centre for the Study of the Networked Image at London South Bank University. She has a long career as a researcher and curator for international organizations and festivals. Dekker publishes regularly in numerous collections and journals and is the editor of several volumes, most recently, *Documentation as Art: Expanded Digital Practices* (Routledge, 2022, with Gabriella Giannachi) and *Curating Digital Art: From Presenting and Collecting Digital Art to Networked Co-Curating* (Valiz, 2021). Her monograph, *Collecting and Conserving Net Art* (Routledge, 2018) is a seminal work in the field of digital art conservation.

Anne Dessing is an architect based in Amsterdam. Since graduating from the Amsterdam Academy of Architecture in 2012, Dessing has been awarded grants from the Creative Industries Fund NL and het Cultuurfonds and has been selected as the 2018–2019 Douglas A. Garofalo Fellow at the UIC School of Architecture, Chicago. Her practice, Studio Anne Dessing, operates at the intersection of art and architecture. She researches architecture through exhibitions, installations, drawings, models, interiors and buildings. Dessing currently teaches at the Academy of Architecture Amsterdam/Rotterdam, the Gerrit Rietveld Academie and the chair of Interiors Buildings Cities at TU Delft.

Mél Hogan is the host of *The Data Fix* podcast and Director of the Environmental Media Lab (EML) in Calgary. She is Associate Professor in the department of Film and Media at Queen's University in Kingston, Ontario. Her research focusses on data centres and communications infrastructure, each understood from within the contexts of planetary catastrophe and collective anxieties about the future.

Michiel van Iersel works as an urbanist, curator and educator at the intersection of the arts, architecture, urban transformation and ecology. He started his career as a curator at several (visual) arts organizations, before initiating his own curatorial practice and (co-)founding, among others, cultural collective Non-fiction and the ongoing research platform Failed Architecture. In 2018/2019 he received a Loeb Fellowship at the Harvard Graduate School of Design. Van Iersel is one of the founding members of Loom, a practice for cultural transformation. He is currently combining his work with Loom with a position as Program Lead of the NEWROPE Chair of Architecture and Urban Transformation at ETH Zürich, where he is also coordinating the Institute of Landscape and Urban Studies. In addition, he teaches, writes and publishes on a broad range of topics, including *Rewriting Architecture – 10+1 Actions for an Adaptive Architecture* (Valiz, 2020).

Max Kortlander is a writer and researcher focussed on participatory democracy, citizen-led governance and human rights in the digital age. He utilizes co-creation and public research methodologies to address real-world problems and keep his work down to earth. His research is always collaborative. He works together with citizens, policymakers, open-source developers and academic researchers through publicly funded projects including DECODE, Public Stack, Digital European Public Spaces, Hollandse Luchten, ACROSS, Urbanite, Cities-4-People and Mobility Urban Values. Kortlander has bachelor degrees in International Relations and Spanish from Bucknell University in Lewisburg, PA and a master degree in Media Studies from the University of Amsterdam.

Mark Minkjan is an architecture critic and urban geographer. He is a co-founding editor of Failed Architecture and has written for publications including *VICE*, *The Guardian* and *The Architectural Review*. He is the co-editor of various books, including *Rewriting Architecture – 10+1 Actions for an Adaptive Architecture* (Valiz, 2020). Minkjan has taught at various design and architecture schools, including the Royal Academy of Art in The Hague, TU Delft and the Academy of Architecture in Amsterdam, and received the Geert Bekaert Award for architecture criticism in 2017. Minkjan is one of the founding members of Loom, a practice for cultural transformation. He has curated programs and exhibitions, including *The Right to Build: Self-build between Dreams and Reality* at the Amsterdam Centre for Architecture in 2019. He curates a public program of live conversations and podcast interviews about alternative building culture and spatial practice at the Rotterdam Academy of Architecture.

Alice Mulder is an illustrator and graphic designer who finds inspiration in travel and daily life. Her illustrations can be recognized by the use of a minimal colour palette and the risograph printing technique. Language plays an important role in her work and she likes to play with different associations that words evoke in her. Mulder's work has been published in international media such as *The Covent Gardener* (UK), *Flow* (Netherlands/France/Germany) and local newspapers. It has been exhibited at (amongst others) Dutch Design Week, Grafixx Festival (Belgium), Chicago Design Museum (USA) and Kaunas Illustration Biennal (Lithuania).

Marina Otero Verzier is an architect, researcher and visiting professor at the Graduate School of Architecture, Planning, and Preservation at Columbia University, New York. Since 2023, she has been a member of the Advisory Committee for Architecture and Design at the Museo Nacional Centro de Arte Reina Sofía. Otero Verzier specializes in the relationships between architecture and digital infrastructures and resources such as lithium that sustain them. In 2022, she received the Harvard Wheelwright Prize for a project on the future of data storage. Between 2020 and 2023, she was the Director of the Master in Social Design at the Design Academy Eindhoven, and from 2015 to 2022, Director of Research at Het Nieuwe Instituut, Rotterdam where she led initiatives focussed on labour, extraction and mental health. Previously, she was Director of Programming for the Global Studio-X Network, Columbia GSAPP. Otero Verzier has curated exhibitions such as *Compulsive Desires: On the Extraction of Lithium and Rebellious Mountains* at the Municipal Gallery of Porto in 2023; *Work, Body, Leisure*, the Dutch Pavilion at the Venice Architecture Biennale in 2018 and *After Belonging*, the Oslo Architecture Triennale in 2016. She has co-edited *Automated Landscapes* (2023), *Lithium: States of Exhaustion* (2021), *A Matter of Data* (2021), *More-than-Human* (2020), *Architecture of Appropriation* (2019) and *Work, Body, Leisure* (2018) among others. Otero Verzier studied at TU Delft, ETSA Madrid and Columbia GSAPP. In 2016, she received her PhD from ETSA Madrid.

Ned Rossiter is Director of Research at the Institute for Culture and Society and Professor of Communication in the School of Humanities and Communication Arts, Western Sydney University. He is author of *Organized Networks: Media Theory, Creative Labour, New Institutions* (2006), *Software, Infrastructure, Labor: A Media Theory of Logistical Nightmares* (2016) and (with Geert Lovink) *Organization after Social Media* (2018).

Niels Schrader is a concept-driven information designer with a fascination for numbers and data. He is founder of the Amsterdam-based design studio Mind Design and member of the AGI – Alliance Graphique Internationale. From 2013 to 2021 he was, together with Roosje Klap, head of the Graphic Design department at the Royal Academy of Art (KABK) in The Hague. Currently he leads the Non Linear Narrative master program at KABK. Next to his design practice Schrader writes regularly for *Grafikmagazin* and *Open! Platform for Art, Culture & the Public Domain* and worked from 2017 to 2020 as a correspondent for *Novum – World of Graphic Design*. In his role as an educator Schrader focusses on social, political and environmental processes driven and influenced by digital technologies. In this context, he has initiated a number of large-scale research projects with partners from outside the academic environment. These institutions include governmental and non-governmental organizations like the Dutch Parliament, Dutch Ministry of Finance, National Archives, National Library of the Netherlands, Free Press Unlimited, Greenpeace, Hivos and Amnesty International.

Jorinde Seijdel is an independent writer, editor and lecturer on subjects concerning art and media in our changing society and the public sphere. She is also editor-in-chief of *Open! Platform for Art, Culture & the Public Domain*. Her articles have appeared in various magazines and publications, including *Spaces for Criticism: Shifts in Contemporary Art Discourses* (Valiz, 2015). She was co-editor of the publication *What's the Use? Constellations of Art, History and Knowledge* (Valiz, 2016). She is theory mentor at the Gerrit Rietveld Academie, Amsterdam and Head of Programme of Studium Generale Rietveld Academie.

Marleen Stikker is a Dutch Internet pioneer, founder of Amsterdam nonprofit organization Waag and author of *Het internet is stuk* (2019). In 1993, she introduced the ground-breaking concept of *De Digitale Stad* (The Digital City), the first virtual community introducing free public access to the Internet in Amsterdam. Stikker strongly adheres to the Maker's Bill of Rights motto: 'If You Can't Open It, You Don't Own It.' She also believes that society needs open technologies that meet societal challenges.

Sander van der Waal is research director of Waag in Amsterdam. He is responsible for its research agenda, working in close collaboration with the four heads of the research groups and leads of the 14 constituent labs. His focus is on ensuring Waag's research contributes to lasting positive change locally in Amsterdam, nationally in the Netherlands and on the European level, in line with Waag's principles of openness, fairness and inclusivity. He maintains and develops a strong network of research partners and represents Waag on stage in debates, events and webinars. Van der Waal has an educational background in computer science and philosophy, and has solid expertise and experience in open-source software development, open data and knowledge, and data infrastructures.

Füsun Türetken is a Professor at the University of the Arts Bremen and University of the Arts Karlsruhe in Germany, and chair of STRÜKTÜR, a Berlin-based arts association, founded in 2024. She holds a doctoral degree from Goldsmiths, University of London. Her PhD was titled 'On the Most Powerful Catalyst on the Planet' and provides a reading of conflict and capital through matter, more precisely, metal. Her academic background lies in architecture, urban studies and visual culture with studies in Germany, the United States and England. In 2008 she was director of the German Pavilion with raumtaktik at the Venice Architecture Biennale. For four years she has been editor and curator with Shrinking Cities in Berlin. Publications include 'The Amalgamated Toxicity of Ground Zero' in *Forensis* (Sternberg, 2014), 'Gold. Alchemic Desire' in *Swiss Psychotropic Gold* (Christoph Merian Verlag, 2020) and 'Neodymium. Superconductive Lifelines for Metallic Monsters' in the journal *UMBAU* (2022). She has shown work in, among other venues, 'Rare Earth, Stories from Below' at STUK Leuven for Artefakt 2018, IMPAKT 2017 and Fiber Festival 2017.

Colophon

Acid Clouds: Mapping Data Centre Topologies

Editors
Niels Schrader
Jorinde Seijdel

***Acid Clouds* Research**
Niels Schrader

***Acid Clouds* Photography**
Roel Backaert

Contributors
Ramon Amaro and Sheena Calvert
Annet Dekker
Mél Hogan
Michiel van Iersel and Mark Minkjan
Max Kortlander, Marleen Stikker
and Sander van der Waal
Marina Otero Verzier
Ned Rossiter
Niels Schrader
Jorinde Seijdel
Füsun Türetken

Illustrations
Anne Dessing
Alice Mulder

Maps and Satellite Images
© Mapbox

Copy Editing
Janine Armin

Programming
Michał Ejdys

Design
Niels Schrader
Mind Design, Amsterdam

Lithography and Printing
robstolk®, Amsterdam

Finishing (Perforation)
Agia, Amsterdam

Binding
Abbringh, Groningen

Paper
Maxioffset 90 gr

Typeface
Graphik by Commercial Type

Publisher
Marcel Witvoet
nai010 publishers, Rotterdam

Thanks to all the contributors to the *Acid Clouds* video interview series published on *www.acidclouds.org*
Ramon Amaro
Michael van den Assem
Evelyn Austin
Walter Hoogland
Stijn Grove
Martin Lloyd
Gerard Rosbach
Lars Ruiter
Marleen Stikker

Special thanks to
Ramon Amaro, Janine Armin, Jacqueline van As, Brecht Bleeker, Gijs Boerwinkel, Patrick Brand, Sheena Calvert, Giulia Cundari, Annet Dekker, Dick van Dijk, Mijke van der Drift, Michał Ejdys, Tjeerd Fonk, Mieke Gerritzen, Yannick van de Graaf, Martijn de Heer, Mél Hogan, Christine van den Horn, Michiel van Iersel, Saskia Jansen, Max Kortlander, Geert Lovink, Mark Minkjan, Marina Otero Verzier, Ned Rossiter, Melissa Scholten, Peter Schrader, Ute Schrader, Marleen Stikker, Füsun Türetken, Alice Twemlow, Wigger Verschoor and Sander van der Waal

***Acid Clouds* is dedicated to the memory of my father, Peter Schrader, who ignited the spark for this research and was always a critical sounding-board for my thoughts, ideas and opinions.**
– Niels Schrader

This publication was made possible by financial support from
- Fonds 21
- het Cultuurfonds
- BPD Cultuurfonds
- J. E. Jurriaanse Stichting

nai010 publishers
Korte Hoogstraat 31
3011 GK Rotterdam
The Netherlands

nai010 publishers is an internationally orientated publisher specialized in developing, producing and distributing books in the fields of architecture, urbanism, art and design.

nai010 books are available internationally at selected bookstores and from the following distribution partners:

North, Central and South America – Artbook | D.A.P., New York, USA, dap@dapinc.com

Rest of the world – Idea Books, Amsterdam, the Netherlands, idea@ideabooks.nl

For general questions, please contact nai010 publishers directly at sales@nai010.com or visit our website *www.nai010.com* for further information.

Printed and bound in the Netherlands.

ISBN 978-94-6208-721-7
NUR 648
BISAC ARC008000, ARC011000

Acid Clouds: Mapping Data Centre Topologies is also available as an e-book:

ISBN 978-94-6208-750-7 (e-book)

FONDS 21 · J.E. Jurriaanse Stichting · bpd cultuurfonds · robstolk drukken printen amsterdam · het cultuurfonds · nai 010

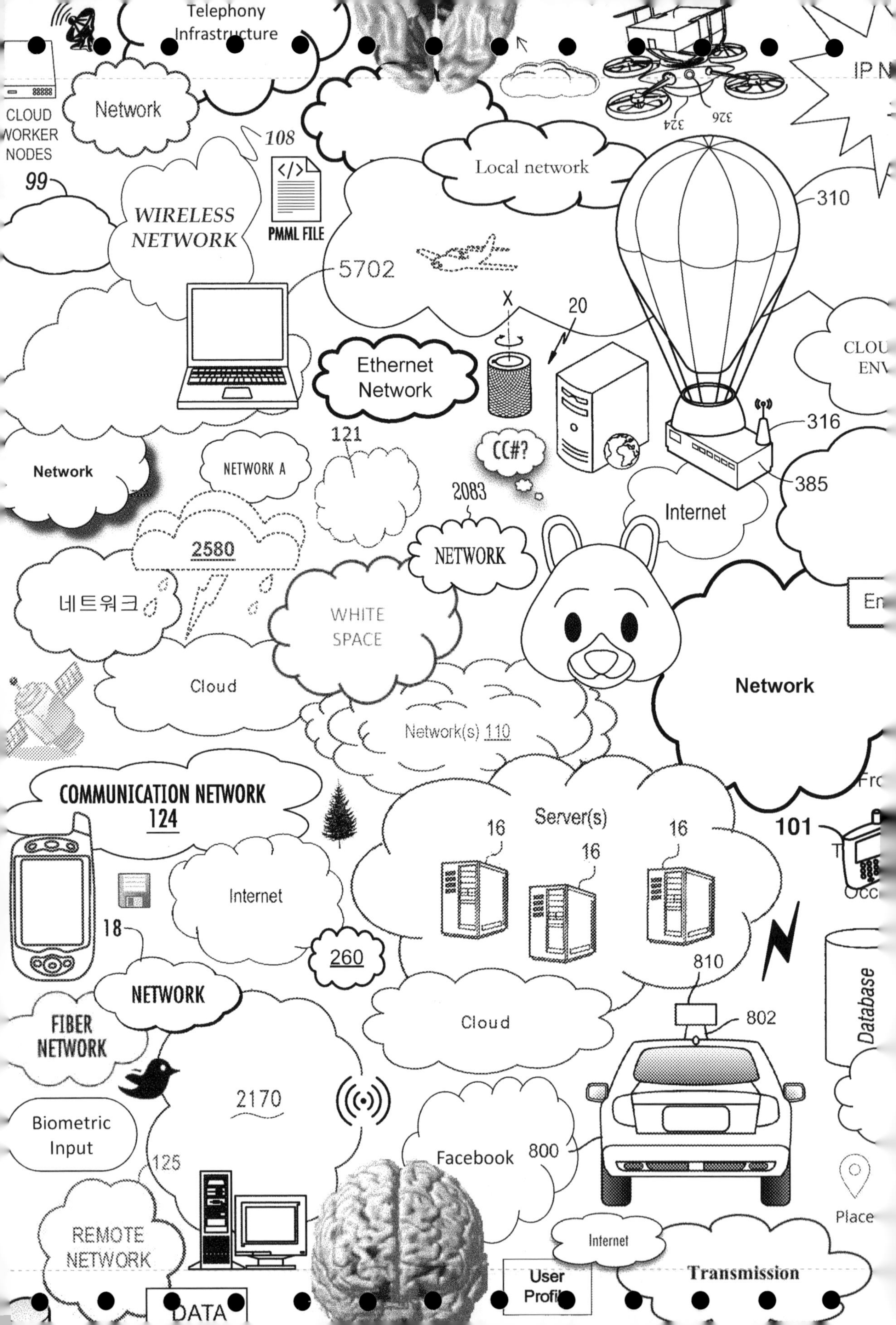

Telephony Infrastructure
CLOUD WORKER NODES
Network
108
99
WIRELESS NETWORK
PMML FILE
Local network
324
326
310
5702
X
20
Ethernet Network
121
CC#?
316
385
Network
NETWORK A
2083
Internet
2580
NETWORK
네트워크
WHITE SPACE
Cloud
Network
Network(s) 110
COMMUNICATION NETWORK 124
Server(s)
16
16
16
101
Internet
18
260
810
NETWORK
802
FIBER NETWORK
Cloud
Database
2170
Biometric Input
125
Facebook
800
REMOTE NETWORK
Place
Internet
Transmission
User Profile
DATA